Lecture Notes in Computer Science

Vol. 219: Advances in Cryptology – EUROCRYPT '85. Proceedings, 1985. Edited by F. Pichler. IX, 281 pages. 1986.

Vol. 220: RIMS Symposia on Software Science and Engineering II. Proceedings, 1983 and 1984. Edited by E. Goto, K. Araki and T. Yuasa. XI, 323 pages. 1986.

Vol. 221: Logic Programming '85. Proceedings, 1985. Edited by E. Wada. IX, 311 pages. 1986.

Vol. 222: Advances in Petri Nets 1985. Edited by G. Rozenberg. VI, 498 pages. 1986.

Vol. 223: Structure in Complexity Theory. Proceedings, 1986. Edited by A. L. Selman. VI, 401 pages. 1986.

Vol. 224: Current Trends in Concurrency. Overviews and Tutorials. Edited by J.W. de Bakker, W.-P. de Roever and G. Rozenberg. XII, 716 pages. 1986.

Vol. 225: Third International Conference on Logic Programming. Proceedings, 1986. Edited by E. Shapiro. IX, 720 pages. 1986.

Vol. 226: Automata, Languages and Programming. Proceedings, 1986. Edited by L. Kott. IX, 474 pages. 1986.

Vol. 227: VLSI Algorithms and Architectures – AWOC 86. Proceedings, 1986. Edited by F. Makedon, K. Mehlhorn, T. Papatheodorou and P. Spirakis. VIII, 328 pages. 1986.

Vol. 228: Applied Algebra, Algorithmics and Error-Correcting Codes. AAECC-2. Proceedings, 1984. Edited by A. Poli. VI, 265 pages. 1986.

Vol. 229: Algebraic Algorithms and Error-Correcting Codes. AAECC-3. Proceedings, 1985. Edited by J. Calmet. VII, 416 pages. 1986.

Vol. 230: 8th International Conference on Automated Deduction. Proceedings, 1986. Edited by J.H. Siekmann. X, 708 pages. 1986.

Vol. 231: R. Hausser, NEWCAT: Parsing Natural Language Using Left-Associative Grammar. II, 540 pages. 1986.

Vol. 232: Fundamentals of Artificial Intelligence. Edited by W. Bibel and Ph. Jorrand. VII, 313 pages. 1986.

Vol. 233: Mathematical Foundations of Computer Science 1986. Proceedings, 1986. Edited by J. Gruska, B. Rovan and J. Wiedermann. IX, 650 pages. 1986.

Vol. 234: Concepts in User Interfaces: A Reference Model for Command and Response Languages. By Members of IFIP Working Group 2.7. Edited by D. Beech. X, 116 pages. 1986.

Vol. 235: Accurate Scientific Computations. Proceedings, 1985. Edited by W.L. Miranker and R.A. Toupin. XIII, 205 pages. 1986.

Vol. 236: TEX for Scientific Documentation. Proceedings, 1986. Edited by J. Désarménien. VI, 204 pages. 1986.

Vol. 237: CONPAR 86. Proceedings, 1986. Edited by W. Händler, D. Haupt, R. Jeltsch, W. Juling and O. Lange. X, 418 pages. 1986.

Vol. 238: L. Naish, Negation and Control in Prolog. IX, 119 pages. 1986.

Vol. 239: Mathematical Foundations of Programming Semantics. Proceedings, 1985. Edited by A. Melton. VI, 395 pages. 1986.

Vol. 240: Category Theory and Computer Programming. Proceedings, 1985. Edited by D. Pitt, S. Abramsky, A. Poigné and D. Rydeheard. VII, 519 pages. 1986.

Vol. 241: Foundations of Software Technology and Theoretical Computer Science. Proceedings, 1986. Edited by K.V. Nori. XII, 519 pages. 1986.

Vol. 242: Combinators and Functional Programming Languages. Proceedings, 1985. Edited by G. Cousineau, P.-L. Curien and B. Robinet. V, 208 pages. 1986.

Vol. 243: ICDT '86. Proceedings, 1986. Edited by G. Ausiello and P. Atzeni. VI, 444 pages. 1986.

Vol. 244: Advanced Programming Environments. Proceedings, 1986. Edited by R. Conradi, T.M. Didriksen and D.H. Wanvik. VII, 604 pages. 1986

Vol. 245: H.F. de Groote, Lectures on the Complexity of Bilinear Problems. V, 135 pages. 1987.

Vol. 246: Graph-Theoretic Concepts in Computer Science. Proceedings, 1986. Edited by G. Tinhofer and G. Schmidt. VII, 307 pages. 1987.

Vol. 247: STACS 87. Proceedings, 1987. Edited by F.J. Brandenburg, G. Vidal-Naquet and M. Wirsing. X, 484 pages. 1987.

Vol. 248: Networking in Open Systems. Proceedings, 1986. Edited by G. Müller and R.P. Blanc. VI, 441 pages. 1987.

Vol. 249: TAPSOFT '87. Volume 1. Proceedings, 1987. Edited by H. Ehrig, R. Kowalski, G. Levi and U. Montanari. XIV, 289 pages. 1987.

Vol. 250: TAPSOFT '87. Volume 2. Proceedings, 1987. Edited by H. Ehrig, R. Kowalski, G. Levi and U. Montanari. XIV, 336 pages. 1987.

Vol. 251: V. Akman, Unobstructed Shortest Paths in Polyhedral Environments. VII, 103 pages. 1987.

Vol. 252: VDM '87. VDM – A Formal Method at Work. Proceedings, 1987. Edited by D. Bjørner, C.B. Jones, M. Mac an Airchinnigh and E.J. Neuhold. IX, 422 pages. 1987.

Vol. 253: J.D. Becker, I. Eisele (Eds.), WOPPLOT 86. Parallel Processing: Logic, Organization, and Technology. Proceedings, 1986. V, 226 pages. 1987.

Vol. 254: Petri Nets: Central Models and Their Properties. Advances in Petri Nets 1986, Part I. Proceedings, 1986. Edited by W. Brauer, W. Reisig and G. Rozenberg. X, 480 pages. 1987.

Vol. 255: Petri Nets: Applications and Relationships to Other Models of Concurrency. Advances in Petri Nets 1986, Part II. Proceedings, 1986. Edited by W. Brauer, W. Reisig and G. Rozenberg. X, 516 pages. 1987.

Vol. 256: Rewriting Techniques and Applications. Proceedings, 1987. Edited by P. Lescanne. VI, 285 pages. 1987.

Vol. 257: Database Machine Performance: Modeling Methodologies and Evaluation Strategies. Edited by F. Cesarini and S. Salza. X, 250 pages. 1987.

Vol. 258: PARLE, Parallel Architectures and Languages Europe. Volume I. Proceedings, 1987. Edited by J.W. de Bakker, A.J. Nijman and P.C. Treleaven. XII, 480 pages. 1987.

Vol. 259: PARLE, Parallel Architectures and Languages Europe. Volume II. Proceedings, 1987. Edited by J.W. de Bakker, A.J. Nijman and P.C. Treleaven. XII, 464 pages. 1987.

Vol. 260: D.C. Luckham, F.W. von Henke, B. Krieg-Brückner, O. Owe, ANNA, A Language for Annotating Ada Programs. V, 143 pages. 1987.

Vol. 261: J. Ch. Freytag, Translating Relational Queries into Iterative Programs. XI, 131 pages. 1987.

Vol. 262: A. Burns, A.M. Lister, A.J. Wellings, A Review of Ada Tasking. VIII, 141 pages. 1987.

Vol. 263: A.M. Odlyzko (Ed.), Advances in Cryptology – CRYPTO '86. Proceedings. XI, 489 pages. 1987.

Vol. 264: E. Wada (Ed.), Logic Programming '86. Proceedings, 1986. VI, 179 pages. 1987.

Vol. 265: K.P. Jantke (Ed.), Analogical and Inductive Inference. Proceedings, 1986. VI, 227 pages. 1987.

Vol. 266: G. Rozenberg (Ed.), Advances in Petri Nets 1987. VI, 451 pages. 1987.

Vol. 267: Th. Ottmann (Ed.), Automata, Languages and Programming. Proceedings, 1987. X, 565 pages. 1987.

Vol. 268: P.M. Pardalos, J.B. Rosen, Constrained Global Optimization: Algorithms and Applications. VII, 143 pages. 1987.

Vol. 269: A. Albrecht, H. Jung, K. Mehlhorn (Eds.), Parallel Algorithms and Architectures. Proceedings, 1987. Approx. 205 pages. 1987.

Lecture Notes in Computer Science

Edited by G. Goos and J. Hartmanis

307

Th. Beth M. Clausen (Eds.)

Applicable Algebra, Error-Correcting Codes, Combinatorics and Computer Algebra

4th International Conference, AAECC-4
Karlsruhe, FRG, September 23–26, 1986
Proceedings

Springer-Verlag

Berlin Heidelberg New York London Paris Tokyo

Editors

Thomas Beth
Michael Clausen
Fakultät für Informatik, Universität Karlsruhe
Haid-und-Neu-Straße 7, D-7500 Karlsruhe 1, FRG

CR Subject Classification (1987): E.4, I.1

ISBN 3-540-19200-X Springer-Verlag Berlin Heidelberg New York
ISBN 0-387-19200-X Springer-Verlag New York Berlin Heidelberg

Library of Congress Cataloging-in-Publication Data. AAECC-4 (1986: Karlsruhe, Germany) Applicable algebra, error-correcting codes, combinatorics and computer algebra. (Lecture notes in computer science; 307) Bibliography: p. Includes index. 1. Error-correcting codes (Information theory)–Congresses. 2. Algebra–Data processing–Congresses. 3. Algorithms–Congresses. I. Beth, Thomas, 1949-. II. Clausen, Michael. III. Title. IV. Series.
QA268.A35 1986 005.7'2 88-12246
ISBN 0-387-19200-X (U.S.)

Printing and binding: Druckhaus Beltz, Hemsbach/Bergstr.
2145/3140-543210

Preface

The present volume contains the proceedings of the conference AAECC-4 held at Karlsruhe, September 23 - 26, 1986. AAECC conferences have been organized regularly since 1983:

AAECC-1: Toulouse 1983
AAECC-2: Toulouse 1984
AAECC-3: Grenoble 1985
AAECC-4: Karlsruhe 1986
AAECC-5: Menorca 1987

The next conferences will be held at Roma (1988), Toulouse (1989) and Yokohama (1990).

The proceedings of the previous conferences were published in 1985 [Discr. Math.**56**,Oct.1985] and 1986 [Springer LNCS 228, 1986] [Springer LNCS 229, 1986]. Quoting from Jacques Calmet´s preface to the AAECC-3 proceedings:

...The main motivation for this series of conferences was to gather researchers in error-correcting codes, applied algebra and algebraic algorithms. The latter topic has been extended to computer algebra in general. Applied algebra must be understood as applied to computer science. After three conferences, it appears that they fill a communication gap. It is thus natural that the AAECC conferences are going to be held annually in different countries. For this reason, a permanent organizing committee has been set up. It consists of Thomas Beth, Jacques Calmet, Anthony C. Hearn, Joos Heintz, Hideki Imai, Heinz Lüneburg, H.F. Mattson Jr. and Alain Poli....

For AAECC-4, 28 papers (plus 4 invited lectures) were presented during the conference. In order to guarantee scientific journal standards, in a second reviewing process which has taken most of the time since the conference each paper was reviewed a second time by a group of at least two independent international experts.

We are grateful to the many people who helped us to organize this conference in such a short length of time. We are especially indebted to Angelika Remmers for her help and support before, during and after the conference. We are also most grateful to the referees whose contributions ensure the significance and quality of the papers published in this volume.

Karlsruhe, February 1988

Thomas Beth
Michael Clausen

GENERAL CHAIRMAN:

T. BETH, University of Karlsruhe, FRG

SCIENTIFIC CHAIRMAN:

H. LÜNEBURG, University of Kaiserslautern, FRG

ORGANIZING COMMITTEE:

T. BETH, University of Karlsruhe, FRG
J. CALMET, LIFIA, Grenoble, France
A.C. HEARN, The Rand Corp., USA
J. HEINTZ, FRG, Argentina
H. IMAI, University of Yokohama, Japan
H. LÜNEBURG, University of Kaiserslautern, FRG
H.F. MATTSON jr., Syracuse University, USA
A. POLI, University P.Sabatier, Toulouse, France

SCIENTIFIC COMMITTEE:

P. CAMION, IRIA, Le Chesnay, France
G.E. COLLINS, Ohio State University, Colombus, USA
B. COURTEAU, Sherbrooke University, Canada
R. LOOS, Universität Karlsruhe and Tübingen after 1.Oct.1986
A. MIOLA, University "La Sapienza", Roma, Italy
W. PLESKEN, Queen Mary College, London, UK
H. ZASSENHAUS, Ohio State University, Colombus, USA

LOCAL ORGANIZATION at the University of Karlsruhe:

M. CLAUSEN
A. FORTENBACHER
D. GOLLMANN
A. REMMERS
V. SPERSCHNEIDER

SESSION CHAIRPERSONS:

T. Beth, J. Calmet, J. Heintz, H. Lüneburg, H. Imai, R. Loos, A. Poli

LIST OF REFEREES:

T. Beth, B. Buchberger, F. Bundschuh, J. Calmet, P. Camion, M.Clausen, G. Collins, B. Courteau, A. Dür, J. von zur Gathen, D. Gollmann, D.Jungnickel, A. Kerber, R. König, W. Küchlin, U. Kulisch, J. van Lint, H. Lüneburg, R. Loos, J. Massey, H.F. Mattson, H. Meyr, A. Miola, H.M. Möller, A. Neumaier, H. Niederreiter, D. Perrin, F. Piper, A. Poli, W. Plesken, A. Schönhage, A. Shamir, J.J. Seidel, N.J.A. Sloane, V. Strassen, W. Trinks

Contents

*A*pplicable *A*lgebra, *E*rror-*C*orrecting *C*odes,
Combinatorics and Computer Algebra

An Introduction to this Volume

Thomas Beth

At a glance the combination of the apparently different fields named in the title seems to be rather artificial. However, the extensive research in these areas during the last decade has shown large intersections and interdependencies between these fields, especially towards several areas of application. These range from *coding theory* and *signal processing* to *software engineering* tools based on *symbolic manipulation* techniques. The use of *data structures* with many *regularities* is the common basis of research in these areas. The regularities lead to algorithm design principles, from which some of the most efficient computational procedures known in the above areas can be developed.

This volume presents a collection of papers all of which can be viewed as variations on the intrinsic relation between algorithm engineering and the regularities of the data structures. We give a short summary of the presented papers. The classification of these papers in one area only, however, seems to be impractical and infeasible. The subsequent survey and introduction will show that it is rather appropriate to orientate the contents of this volume in accordance with the general aims of the AAECC-conference series in form of the following pentagram:

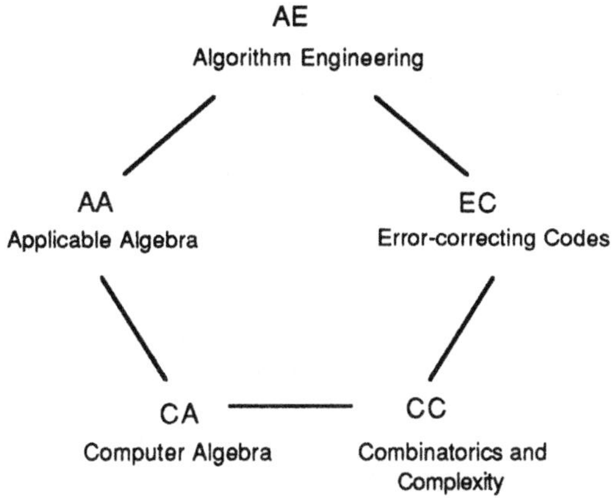

The paper of *Bazard* and *Huguet* shows the close relation of single error-correcting codes to triangular graphs by which a rather regular combinatorial structure is defined in order to derive with methods of linear algebra the stated results. The connection between regularity of codes and their presentation by tools of linear algebra is given in the work by *Rifa* and *Huguet*

on the characterization of codes by so-called association schemes. The importance of these combinatorial structures has been pointed out by *Delsarte* [9] and has been used since to determine weight distribution of codes as is also done in the paper by *Dür*, who determines the weight distributions of some special Goppa codes. This question is interesting and important for at least two reasons: on the one hand weight distributions of codes allow for a rather direct calculation of residual error probabilities in additive symmetric channels. On the other hand weight distributions can be used to determine possible regularities in the data structures presented by such codes.

Another example for the close interrelation between the underlying data structures and the efficiency of the required algorithms is given in the paper by *Ericson* on the Blokh-Zyablov-decoding algorithm. This algorithm has raised quite some interest throughout the last few years because it seems to be one of the most efficient decoding features applying the ideas of concatenated codes in a very efficient and clear way. Concatenated codes [11] are nothing but record-like data structures which allow for refined subrecords.

The paper by *Garcia-Ugalde* and *Morelos-Zaragoza* yields an implementation of a serial Viterbi decoding algorithm. The Viterbi decoding algorithm is of high practical significance, it is an example of dynamic programming on a data structure which initially is to be considered as a "hypercube" over GF(2). For further studies, cf. MacWilliams/Sloane [12].

The note by *Griera* relates the algebraic geometric structure of codes to the combinatorial interpretation via its weight distribution.

Similar in spirit, but quite different in its tools is the paper presented by *Opolka* on the embedding of self-dual codes into projective space which demonstrates the close interrelation between coding theory and representation theory. At this point it should be mentioned how important the tool of group theory is in dealing with these kinds of regular algorithms , cf. Beth [1]. The same observation applies to the paper by *Poli* discussing the construction of self-dual multicirculant codes. The structure in these codes which allow for a large set of automorphisms needs tools from modular representation theory and polynomial algebra in several variables.

Shokrollahi uses algebraic geometric tools in his paper dealing with the construction of codes on algebraic curves. This vastly extends the concept of cyclic codes and Goppa codes as has been done in the area of coding theory for the last two decades, cf. van Lint [11]. His work shows the need for more complicated data structures in order to approach the bounds for so-called good codes as had been pointed out by Tsfasman, Zenk and Vladut some years ago [2]. *Thiong-Ly* uses in his paper on the symmetries of conventional Goppa codes classical polynomial arithmetic methods. The last two mentioned papers set forth some trends in algebraic coding theory.

The paper by *Opolka* on the connections between finite Fourier transforms and modular forms points out another aspect of the importance of classical algebraic tools in dealing with the questions of coding theory from a higher point of view. The relation to Fourier transforms shows again the use of methods from group representation theory.

The paper by *Pittaluga* and *Strickland* is concerned with the computation of the (modular) irreducible K-representations of the special linear groups SL(n,K) over an infinite field K. The contribution by *Massazza, Mauri, Righi* and *Torelli* presents a symbolic manipulation system for the solution of counting problems in terms of generating functions. This is an interesting example of the importance of computer algebra systems for the research in other fields. Some systems like MACSYMA [3], "Scratchpad" [4] and SAC-2 [5] are provided for such symbolic manipulations. The tools of software engineering applied here are closely related to the algebraic combinatorial methods by which the problem specification takes place. Similar aspects are dealt with in the paper by *Mascari* and *Miola* on integration of numeric and algebraic computations. Designing such computer algebra systems requires intensive research in the area of abstract data types and in order to automatize the problem specification a mechanism for type inference is needed in future systems. The paper by *Calmet, Comon* and *Lugiez* presents a short account of an approach to this problem. The design of scientific workstations is also related to the availability of graphical tools and their connection to the underlying computer algebra system. This question is addressed by *Bittencourt* in his paper about the integration of graphical tools in a computer algebra system.

In this context the paper by *Poli* and *Gennero* shows a very efficient software program for factorization of polynomials over large finite fields. A reduction relation together with a Gröbner basis provides a data structure for computing with ideals in polynomial rings. *Weispfenning* gives some bounds on the Buchberger algorithm computing a Gröbner basis and *Mora* extends some results on Gröbner bases to the non-commutative case.

Trace languages, closely related to the combinatorial theory of rearrangements, are the subject of the paper by *Bertoni, Goldwurm* and *Sabadini*. The authors study the time complexity of a class of algorithmic problems on trace languages both in the worst and in the average case.

This summary has given an overview of the different topics approached in papers in this volume. The combination of these papers shows one of the main lines of research in the area of applicable algebra and computer algebra exemplifying the most interesting problem of how to combine software engineering tools and algorithm engineering tools in order to give automatical specification and solution methods into the hand of users who are working in application areas in which regular data structures and efficient algorithms are of large importance.

References:

[1] T. Beth: Verfahren der schnellen Fourier-Transformation, Teubner 1984

[2] M.A. Tsfasman; S.G. Vladut; T. Zink: Modular Curves , Shimura Curves, and Goppa Codes better than the Varshamov-Gilbert-Bound, Math. Nachr. 104, 13-28, 1982

[3] Pavelle, R.; Wang, P.S.: MACSYMA from F to G, J. Symb. Comp. (1985),1, 69-100

[4] R.D. Jenks; R.S. Sutor; S.M. Watt: SCRATCHPAD II: An abstract data type system for mathematical computation, in: R.Janßen (ed.): Trends in Computer Algebra, International Symposium Bad Neuenahr, Proceedings LNCS 296, 1988

[5] SAC-2: by G. Collins, Ohio State University; R. Loos, University of Tübingen

[6] Proceedings AAECC-1: Discrete Mathematics, Vol.56, No. 2-3, Oct. 1985

[7] Proceedings AAECC-2: Applied Algebra, Algorithmics and Error-Correcting Codes, (ed. by A. Poli), Springer LNCS 228, 1986

[8] Proceedings AAECC-3: Algebraic Algorithms and Error-Correcting Codes (ed. by J. Calmet), Springer LNCS 229, 1986

[9] P. Delsarte: An Algebraic Approach to Coding Theory, Philips Res.Rep.Supp. 10, 1973

[10] N.J.A. Sloane: Error-Correcting Codes and Invariant Theory, A.M.M, Feb.1977, 82-107

[11] J.H. van Lint: Introduction to Coding Theory, GTM 86, Springer 1982

[12] F.J.MacWilliams; N.J.A. Sloane: The Theory of Error-Correcting Codes, North Holland 1979

1. Graphs: definitions and basic properties

A **graph** G(V,E) is formed by a vertex-set V and an edge-set E, where every edge is a subset of cardinality 2 of V. The graph is **regular** if every vertex appears in the same number of edges.

Let G(V,E) be a connected graph with $|V|=n$ and $|E|=m$. A **path** is a sequence of adjacent edges, and a **tree** of G is any subset T of E with $|T|=n-1$ that establishes one and only one path between every pair of vertices in G.

For a tree T of G any $e_i \in E / e_i \notin T$ is called a **chord** of T.

Let $Y_{n \times m}$ be the **(0-1)-incidence-matrix** of G where

$$Y(i,j)= \left| \begin{array}{l} 1 \text{ if } e_j \text{ links } v_i \\ 0 \text{ otherwise.} \end{array} \right.$$

A **circuit** is a path that begins and finishes at the same vertex. The $m-(n-1)$ circuits obtained by the addition of any chord to T are called **fundamental circuits**. Let C_f be the **(0-1)-fundamental circuit matrix** with m columns and $m-(n-1)$ rows where

$$C_f(i,j)= \left| \begin{array}{l} 1 \text{ if } e_j \in \text{ circuit } i \\ 0 \text{ otherwise.} \end{array} \right.$$

A **cut-set** is any subset U of edges whose deletion disconnects the graph. U is called minimal if $|U|$ is minimum.

The n-1 minimal cut-sets with respect to T that contain one and only one edge of T are called **fundamental cut-sets**. Let T_f be the **(0-1)-fundamental cut-set matrix** with m columns and n-1 rows where

$$T_f(i,j)= \left| \begin{array}{l} 1 \text{ if } e_j \in \text{ cut-set } i \\ 0 \text{ otherwise.} \end{array} \right.$$

Now we can consider the matrix Y as $Y=(Y_1|Y_2)$ where the columns of Y_1 are put as the chords of a fixed tree.

FROM T(m) TRIANGULAR GRAPHS TO SINGLE-ERROR-CORRECTING CODES

J.M.Basart and L.Huguet
Dpt. d'Informàtica, Fac. de Ciències, Univ. Autònoma de Barcelona, 08193 Bellaterra, Catalonia

Abstract

We present a way to construct an infinite family of linear codes, with d=3, from a particular type of strongly regular graph.

Using the incidence matrix and taking a tree of any nondirected and connected graph G(V,E) we can obtain the matrix of fundamental circuits and the one for fundamental cuts. These are respectively the generator and parity check matrices of a linear code.

We are interested in the construction of triangular strongly regular graphs, named T(m). From T(m) and some results on the girth and the valency of a strongly regular graph G, for any integer $m \geq 4$, we obtain a linear code C(T(m)) with parameters: $n=(m(m-1)(m-2))/2$, $k=(m(m-1)(m-3)+2)/2$, $d=3$ and $A_i=A_{n-i}$, A_i being the number of codewords of weight i in the code.

Moreover we give some properties of its codewords set and its orthogonal one. Finally, an alternative method using lattice graphs L_2 is proposed.

Moreover we can write C_f and T_f as

$$C_f=(I\,|\,\overbrace{Y_1^t(Y_2^t)^{-1}}^{C_g}) \qquad \text{and} \qquad T_f=(C_g^t\,|\,I) \qquad \text{(ref. 2)}$$

Obviously C_f can be considered as the systematic generator matrix of a binary linear code C, and consequently T_f will be its parity-check matrix.

This remark is the basis for the following considerations.

2. Strongly regular graphs

A **strongly regular graph** is a regular graph neither complete -that is it has not all the possible edges- nor null. Its parameters are N,A,C and D where

N is $|V|$,

A is the valency or common degree of each vertex and

C and D are respectively the number of vertices adjacent to v_i and v_j according to v_i and v_j are adjacent or non-adjacent.

Examples:

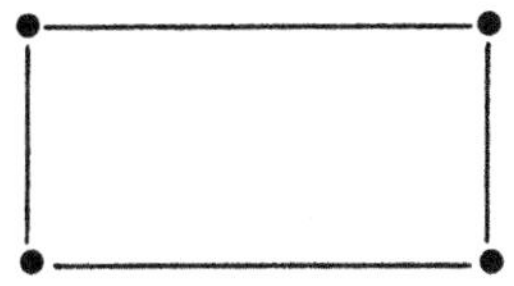

N=4 A=2 C=0 D=2

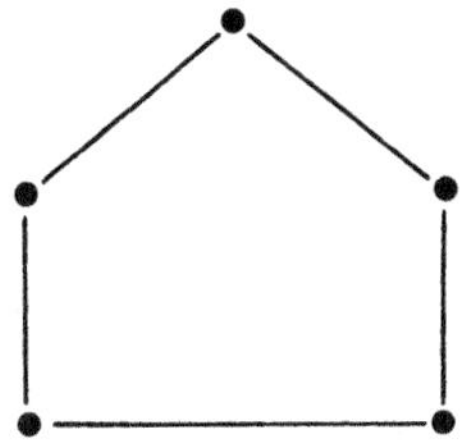

N=5 A=2 C=0 D=1

The **girth** of G is the number of edges in the shortest circuit of G.

Every strongly regular graph with valency greater than 1 has:

girth=3 if C>0

girth=4 if C=0 and D>1 (ref. 1)

girth=5 if C=0 and D=1.

We are interested in a special type of strongly regular graph, the so called **triangular graphs** T(m).

In these graphs T(m) -where $m \geq 4$- the vertex-set is the set of all 2-subsets of an m-set, where two different vertices make an edge if and only if they are not disjoint sets. Their parameters are:

N=S(m,2), A=2(m-2), C=m-2 and D=4.

(where S(p,q) means the number of q-subsets in a p-set.)

So independently of its m, T(m) has girth=3.

Example:

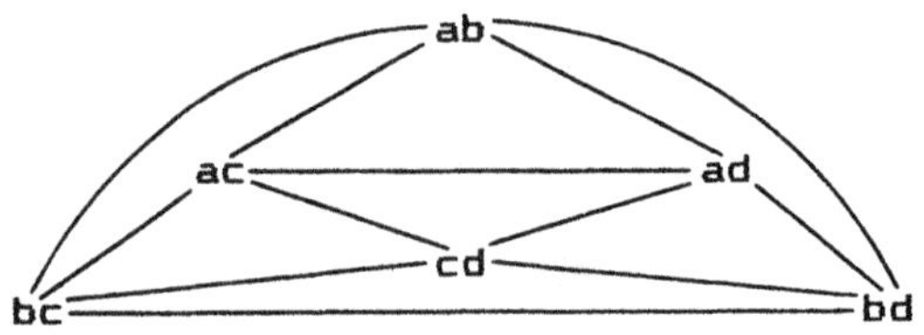

Obtained taking m=4 and the set {a,b,c,d}.
T(4) has N=6, A=D=4 and C=2.

3. Main results

Using the usual definition of length (n), dimension (k), and minimum distance (d) in a linear code C we establish that:

(i) The C(T(m)) code obtained using T(m) has parameters:

n=m(m-1)(m-2)/2, k=(m(m-1)(m-3)+2)/2 and d=3.

These values are directly found from the T(m) parameters since n=(N A)/2, k=(N A)/2 - (N-1) and d -the minimum weight- is the number of edges in its shortest circuit.

(ii) C(T(m)) contains the "all-ones" codeword.

This fact is due to an old theorem (Euler) about Eulerian circuits in a graph G -circuits which pass once and only one through every edge-. This theorem establishes that an Eulerian circuit exists in G if and only if all its vertices have even degree, and this is our case in T(m).

(iii) Let A_i be the number of codewords of weight i in C(T(m));

then $A_3=m(m-1)(m-2)^2/6$.

Really, to start to construct a circuit of 3 edges we can choose among N=S(m,2) vertices. Let v_i be one of them. Now we have A=2(m-2) options for the next v_j, and when we have it, there are 1+(m-3) ways to select the last v_k so that an edge (v_k,v_i) exists.

We must divide this amount by 2 since v_i---v_j---v_k and v_i---v_k---v_j are the same circuit, and finally to divide by 3 because at the end each circuit will be counted three times, one for every first vertex.

Thus we have

$$A_3 = \frac{S(m,2)\,\dfrac{2(m-2)(1+m-3)}{2}}{3} = \frac{m(m-1)(m-2)^2}{6}$$

(iv) $C(T(m))^{\perp}$ has d=2(m-2).

Since every vertex is in A edges, to make a cut-set we shall to take out at least all the A=2(m-2) edges in one vertex. In fact if we want to make a cut-set with r≥1 vertices in one side and s=2 vertices in the other one we shall need:

2A − 2	edges if both vertices are linked by an edge and
2A	edges otherwise.

If s=3 we shall need:

$3A - 6$	edges at minimum, when the three vertices are linked by three edges and
$3A$	edges at maximum, when they have not common edges.

And so on for $s \geq 4$. In every case the number of required edges will be greater than A.

(v) $C(T(m))^{\perp}$ has $A_i = 0$ for every odd i.

First we must remind a well-known result on graph theory:

The number of odd-degree vertices in any graph is always even.

To prove (v) is equivalent to explain that all the fundamental cut-sets in T(m) will use an even number of edges.

Really, to get a fundamental cut-set involves share the graph in two components. Let G_1 be one of them.

If G_1 has only one vertex our statement is true since the valency is even. If G_1 has more than one vertex its number of odd degree vertices will be even, thus the total number of taken edges from these vertices will be also even. Adding to it the number of edges taken from the even degree vertices in G_1 we obtain again an even number.

(vi) The number of minimum weight codewords in $C(T(m))^{\perp}$ is

$A^{\perp}_{2(m-2)} = S(m,2)$.

Of course, because there will be the same number of codewords of minimum weight as there are fundamental cut-sets, that is N, one for each vertex.

4. Example

For the triangular graph T(4) we find that the code C(T(4)) has parameters: n=12, k=7 and d=3 with weight distribution

i	0,12	3,9	4,8	5,7	6	otherwise
A_i	1	8	15	24	32	0

and its orthogonal code $C(T(4))^{\perp}$ has

i	4	6	8	otherwise
$A_i^{\perp}$	6	16	9	0

Remark: We can obtain similar results using another family of strongly regular graphs, the so called **lattice graphs**, $L_2(m)$.

These graphs have the set V formed by all the elements of the SxS cartesian product of a set S with m≥2 points.

Two different vertices will be in the same edge if and only if they have one point in the same position. Its parameters are: $N=m^2$, $A=2(m-1)$, $C=m-2$ and $D=2$.

5. Further research

In this paper we have applied some known ideas -interested reader should see ref. 3,4 and 5- and developed other graph concepts into the construction of linear codes. But this is only an approach, more research ought to be done:

(a) To find more -all, perhaps is also possible- coefficients in the weight distribution of C(T(m)) and $C(T(m))^{\perp}$. Probably new ideas will be required.

(b) Looking for the T(m) covering radius.

(c) To relate the C(T(m)) parameters to the parameters of other codes obtaining from the result some appropriate applications.

Acknowledgements

The autors are very grateful to the referees for their comments and suggestions, which have really improved our initial work.

References

(1) "Graph theory, coding theory and block designs"
P.J. Cameron and J.H. Van Lint
Cambridge University Press, Cambridge 1975.

(2) "Graph theory"
W. Mayeda
John Wiley and Sons, New York 1972.

(3) "Error Correcting Codes"
W.W. Peterson and E.J. Weldon Jr.
MIT Press, Cambridge 1972.

(4) "Cut-Set Matrices and Linear Codes"
S.L. Hakimi and H. Frank
IEEE Trans. on Inf. Theory 11 (1965) pp. 457-458.

(5) Bobrow and Hakimi
IEEE Trans. on Inf. Theory 17 (1971) pp. 215-221.

(6) "Configuracions i grafs en la teoria de codis"
J.M. Basart
Dissertation for 1st. degree.
Univ. Autònoma de Barcelona, Bellaterra 1985.

Integration of Graphical Tools in a Computer Algebra System

Guilherme BITTENCOURT
LIFIA - BP 68
38402 Saint Martin d'Hères cedex, France

Abstract

This project is embedded in the framework of the development of a programming environment integrating symbolic, numeric and graphical tools, in an Artificial Intelligence approach. To carry out the interfacing between a Computer Algebra System and graphical tools the GKS graphical standard has been chosen. Franz-LISP was selected as the host language to implement the system.

Once the graphical kernel is implemented, the interface modules may be written by making use of the GKS primitives. These modules are of two types : the user's interface modules, which display mathematical expressions in human adapted forms, and the symbolic-graphical interface, which allows the plotting of mathematical functions.

Presently the graphical kernel is in test phase, and the interface modules are being specified. Some prototypes of the interface modules are also ready for testing.

1. Introduction

This paper describes the implementation of some modules of a system presently under development (Calmet and Lugiez, 1987). This system intends to integrate graphical, numerical and symbolic capabilities in a scientific computing system.

The modules described here correspond to the graphical kernel of the system and to the symbolic-graphical interface. Computer Algebra systems dispose, in general, of some graphical capabilities, but they are either very primitive or dependent on the graphical terminal type. For instance REDUCE's (Hearn, 1985) only graphical capability is the two-dimensions exponent output. MACSYMA (Martin and Fateman, 1981) disposes of some mathematical sign like summation and integral as well as fraction lines but these symbols are constructed with ordinary characters and not with graphical primitives. A simple plot fuction is also available in the MACSYMA system but the graph is created using ordinary characters and not graphical capabilities. The Maple system (Char et al., 1985) also disposes of a two-dimensions output facility but always using ordinary characters. Several text processing systems (Lamport, 1986) allow sophisticated mathematical output but these systems are typically not available on line. Using a graphical standard kernel system to control the graphical devices and implementing graphical capabilities in a Computer Algebra system relying on the standard higher level graphical primitives allow both terminal independence and powerfull graphical capabilities. Finally, a proposition for the architecture of an intelligent interface able to treat part of the problem of graphical-numerical-symbolic intercommunication is also presented.

In section 2, the implemented graphical system, based on the GKS (Graphical Kernel System) standard, is described. In section 3, four applications implemented by making use of the graphical kernel are presented : usual textbook notations, tree structure visualization of mathematical expressions and one and two variables function plotting. In section 4, an architecture designed to treat the graphical-numerical-symbolic interfacing problem in an unified way is proposed. Finally, in section 5, some conclusions regarding the implementations and the directions of future work are presented.

2. The Graphical Kernel System

Since 1982, GKS (Hopgood et al., 1983) is adopted as an International Standard in Computer Graphics. Two main reasons explain why this standard has been selected for our project. The first one is to enable an easy integration into the system of modules of graphical algorithms designed accordingly to the GKS standard. The second one is that it insures terminal independence in input and output as well.

2.1. Description

Workstation is a central concept in GKS. It stands for a set of output and/or input devices treated by the system in an uniform way. During the initialization of the GKS system a data structure relating physical devices characteristics to workstation types is created, the contents of this data structure depends on the hardware environment. Once a workstation type has been declared to the system the user may open one workstation of this type and assign a name to it. In the application programs all references to workstations are made using only the assigned names, which garanties total independence between applications and physical device characteristics.

Three kind of coordinates systems are present in GKS : world coordinates (WC), normalized device coordinates (NDC) and workstation coordinates (WSC). The user may declare as many WC systems as needed, the graphical data will allways refer to one of these systems. There is only one NDC system and it corresponds to a unity side size square. There are as many WSC systems as workstations opened. Every output in GKS involves two coordinates transformations : WC to NDC and NDC to WSC. The window and viewport sizes of both transformations are under user control.

The graphical data storage in GKS is accomplished using segments. A segment is a set of output and/or control primitives stored under a name. All graphical data are stored using the NDC system, and once a segment is created it can be transformed in three different ways : scale change, rotation and translation. Besides normal workstation dependent segment storage, GKS allows a workstation independent segment storage. This permits to a segment created interactively in one workstation to have its final version copied into another workstation.

Four output primitives are proposed in GKS : Polymarker, Polyline, Fillarea and Text. The first three primitives receive as parameter a list of point coordinates and generate, respectively, one special mark at each point coordinate, the line segments linking the points and the filled polygon defined by the points. The fourth primitive allows someone to write strings of characters in given positions of the screen. All four output primitives have attributes which control the characteristics of the displayed image : kind of marks, lines or filling patterns, font and size of characters, etc. These attributes are also under user control.

The GKS standard proposes also six input primitives. These primitives correspond to virtual input devices and the physical counterpart associated to each primitive may vary from one installation to another one. The primitives and their functions are : LOCATOR returns the coordinates of one point, STROKE returns a list of coordinates of a set of points, PICK returns the name of the segment where the pointed part of the picture is stored, CHOICE returns a natural number, VALUATOR returns a floating-point number and STRING returns a string of characters.

Three kind of interaction modes are proposed in GKS : REQUEST mode, SAMPLE mode and EVENT mode. The REQUEST mode corresponds to the usual READ instructions of the programming languages, that is, the application program requests an input and waits until the input process obtains the input data and sends it back. When the application program receives the input data it continues to execute. In the two other modes the application program and the input process are both active at the same time. In the SAMPLE mode the application program has the control and can obtain at each moment the last input data received. In the EVENT mode the input process has the control, each input data received is stored in a queue which the application program should access. So, each data entered by the operator will necessarilly be received by the application program and in the same order that it has been entered.

As a standard, GKS must include systems of different sizes and powers. To implement this, nine levels of GKS implementations are proposed : three levels for output capabilities (0, 1 and 2) and three levels for input capabilities (a, b and c).

2.2. Implementation

The GKS implementation described here is of level **2b**, which means, complete output capabilities and input in REQUEST mode only. The SAMPLE and EVENT modes are used mainly in applications that require animation. For applications which need only static interaction, as in the case of the proposed interface, these modes are not essential. Besides, the inclusion of these modes would present a technical problem if the language chosen to implement the system does not include interrupt facilities. These facilities would have to be implemented in another language and then integrated into the system.

The Franz-LISP (Foderaro and Sklower, 1982) language was selected to implement the system. This choice is due to several reasons : LISP is the base language of most Computer Algebra systems, LISP is well adapted to the implementation of an intelligent interface and, finally, LISP presents many useful capabilities (modularity, dynamic memory, property lists, efficient list manipulation, etc).

The data structure chosen to represent the graphical data is the LISP association lists (A-lists), where each pair in the list stands for one point coordinates. Thus, all manipulations on graphical data are accomplished using the A-lists manipulation functions. To store control data the LISP property lists (P-lists) are used. System general data are stored in the P-list associated to the symbol **gks**, the data corresponding to workstations, segments and normalization transformations are stored in the P-lists associated to the symbols chosen by the user to label these entities.

The graphical system is composed of two kind of functions : central system functions and terminal control functions. The central system functions are grouped into modules following closely the decomposition proposed by Krüger et al. (1985). The interdependency of these modules are presented in figure 1, where an arrow from a module **A** to a module **B**, means that some function in module **A** calls some function in module **B**. The terminal control functions are grouped into special files, one for each workstation type available in the actual installation. All communication between the two kind of functions are made through one specific module of the central system (**outputs** module).

To allow the system to recognize a new kind of terminal it is necessary to write ten control functions specific for that kind of terminal (one for initialization, one to clear the display surface, four to implement the output primitives and four to implement the input primitives). These control functions depend on the terminal type but in all cases they are very simple to write.

The system accepts three categories of terminals : pixel access terminals, vector terminals and raster terminals. Terminals of the first category are able only to set on or to set off one pixel of the screen, so even a line segment must be constructed by software. Terminals of the second category have the line segment drawing capability, but to fill a defined area of the screen it is necessary to write a special program able to determine the line segments which, when displayed, will fill the defined area. Finally, raster terminals have vector and fill area capabilities. All algorithms (Foley and Van Dan, 1982) to implement GKS primitives using just pixel access capability are then present in the system, but when a specific terminal has some capability it is used instead of the internal algorithm. Some of these internal algorithms, mainly the area filling algorithm, are very time consuming and typically are implemented in hardware. So, in our system, the filling area primitive in a pixel access only terminal will take necessarilly some time. This technical difficulty can be reduced by implementing these algorithms in assembler language and integrating them into the system. This integration is easy to do using the Liszt Compiler (Foderaro and Sklower, 1982) capabilities.

The syntax proposed in the GKS standard has been modified in some details in the present implementation. All arrays of numbers have been replaced by A-lists and all variables representing limits of any kind (number of points, number of workstations, number of segments, etc) have been deleted. Also some GKS primitives are supposed to return some values in their parameters, in the present implementation these values are returned by the primitives themselves, making use of the normal functional interpretation of LISP routines. In the appendix, the retained syntax of the GKS primitives and a brief description of their functions are presented. This description is useful in the sense that the reader can have an idea of the form of GKS primitives and how they can be used in graphical procedures.

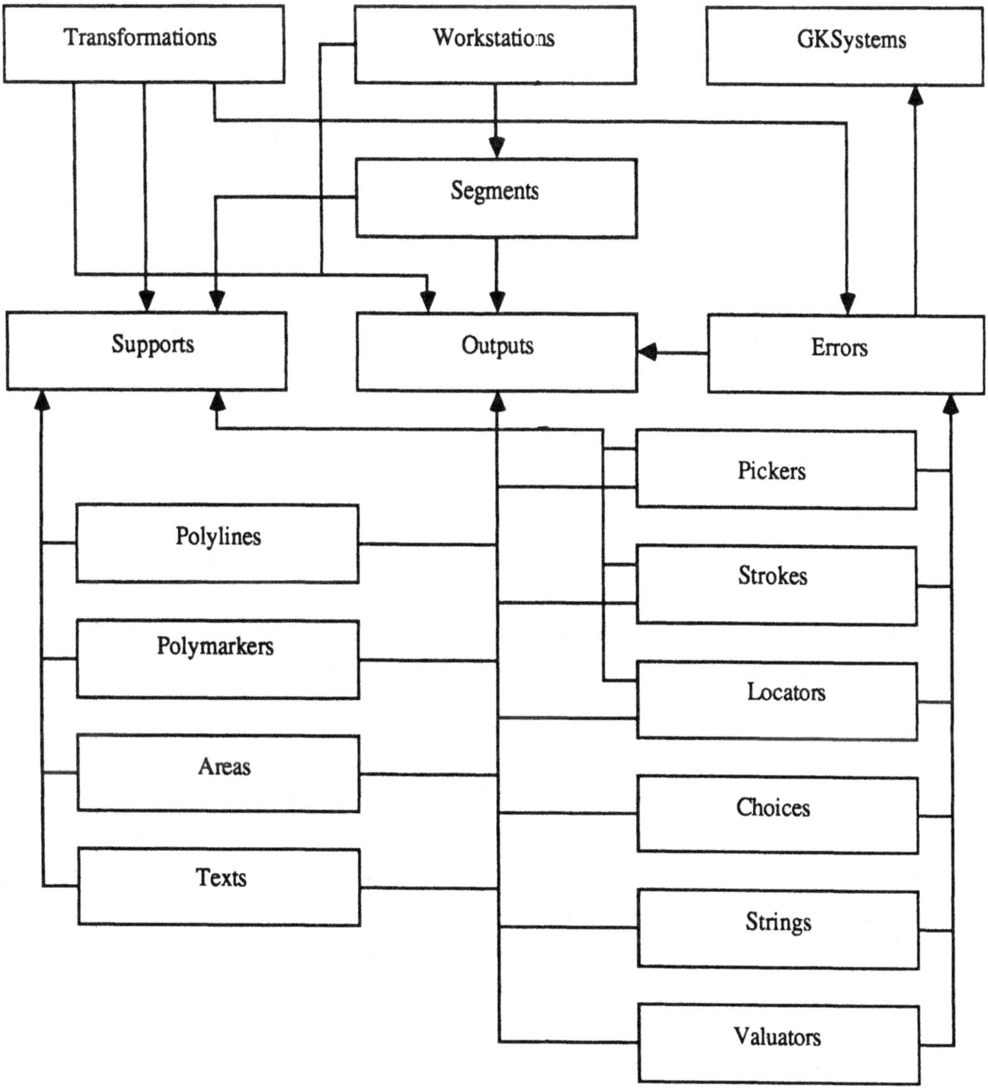

Fig. 1 - Dependency of the Modules

3. Applications

In this section four GKS applications are described. These applications intend to demonstrate the possibility of implementing, using the GKS primitives, the graphical tools needed in a symbolic-graphical interface. Each of these applications is a LISP function with special parameters, which will be called by the future intelligent interface when needed. The way these parameters will be chosen and how the functions will be called are described in section 4.

3.1. Mathematical Expressions Display

Two applications were implemented to allow the visualization of mathematical expressions : a textbook notation and a tree structure representation. The textbook notation application is based on the work of Martin (1971). The input to this function is a mathematical expression coded in a LISP-like syntax. The output presented to the user is that expression written in the normal mathematical notation, including integral and sumation symbols, fraction lines, etc. The image is stored in a segment where each subexpression is individually identified, allowing the user to obtain the symbolic expression, in LISP syntax, of a part of the expression which he has pointed to. This application can be easly integrated in an expression editor. No interrupt facility is needed for this kind of interaction where the system shows an expression and waits for an answer from the user.

The tree structure display application has the same form of input as the previous application. Its output consists of a tree representation of the input expression, where each node corresponds to an operator and the descendends of that node to the corresponding operands. Leaves are either constants or variables. As in the previous application, the image is stored in a segment, allowing the user to select one node by pointing to it and to obtain as a result the symbolic expression of the sub-tree with root in the chosen node.

The fact that the images are stored in segments allows to display the same image in another workstation or the same image transformed in size or position without any extra computation.

3.2. Function Plotting

Two function plotting applications have been developed : one variable and two variables function plotting. The input to the one variable function plotting application is a list of mathematical expressions in the LISP syntax, the name of the independent variable and the plotting limits. The application generate an image where all the functions in the input list appear. The zero line of the ordinate is also represented. Each curve appears with a different line style (if the terminal is able to draw the needed line styles). The image is stored in a segment. This allows the user to have the same image translated, possibly in a different scale, whitout extra calculations of the curve values. The user can also choose one curve by pointing to it and obtains as result the symbolic expression of that curve. As in the mathematical expression display application the system outputs the curves and waits for the user answer without need of interrupt facilities.

The two variables function plotting application receives as input one function expression, the names of the independent variables, their respective limits and the desired perspective. As output the application generates the three dimensional representation of the surface. This application is based on the algorithms developed by Williamson (1972) and Wright (1973). The graph is composed of two series of equally spaced curves representing the surface, curves in one series are parallel to the **x** axis and curves in the other series are parallel to the **y** axis, the algorithm determines the hidden parts of each curve. This picture is also stored in a segment allowing to display the same image in a different scale whitout any extra computation.

4. The Intelligent Interface

Symbolic, graphical and numerical informations are intrinsicaly different, and the efficient transformation of data and methods from one of these fields to another is highly dependend on the specific case and is, in general, an open problem. The interface proposed here is based on a knowledge representation formalism (Fikes and Kehler, 1985) allowing storage, classification and manipulation of specific knowledge of each of these three fields and special cases of transformations from one field into another one. The retained architecture is similar to that of an Expert System. It consists of a base of rules, a working memory module and an inference engine.

The rules are represented by frames classified in a hierarchical form. Each rule is composed by a set of slots specifying the condition and the action part, the set of successor rules and the set of parent rules. Besides these slots, some rules can have special slots to store information specific to the manipulation of the objects to which the rule refers to, like operators, transformations, predicate functions, etc. Other slots, like an external representation of the rule to be presented to the user, can be added to the rule. There are properties associated to the slots that specify the format of the contents of the slot and the way these contents should be interpreted. The slots and their properties are inherited by each rule from its parent rules. This architecture allows that rules refering to different types of knowledge have different formats and interpretations of the condition and action parts. It allows also the association of procedures to the rules in a flexible way, this feature is very important to manipulate mathematical knowledge. The hierarchical classification of rules also enables a set of rules to be loaded only when it is necessary to treat the problem at hand.

The working memory module stores trace information, control variables values and expressions representations. The inference engine controls the whole process by selecting and applying rules and by creating and modifying the working memory data.

This architecture has two main advantages : flexibility and generality. The base of rules may be changed at any time and new applications may be added to the system simply by adding new sets of rules. Moreover, many classes of problems may be treated by this model of architecture: it is enough to write the appropriate rules and to add this set of rules to the base of rules. Another advantage is that the interfacing mechanism is independent of both the Computer Algebra kernel and of the graphical kernel thus allowing to use these modules in an independent way when interfacing facilities are not needed, like in batch processing for example.

In the case of the symbolic-graphical interface the following sets of rules must be created : one set of rules for each kind of mathematical expression notation wanted (the system can support as many specific notations as needed), one set for one variable function plotting control, one set for two variables function plotting control and finally one higher node in the hierarchy with the function to activate the other sets of rules when needed. The graphical functions described in section 3, when integrated into the system, will be activated by the action part of some rule included in the above sets of rules, also their parameters will have their values determined by specific rules.

Other sets of rules may be constructed to treat problems like symbolic-numerical interfacing and solution of specific mathematical problems (differential equations, Laplace transforms, etc). Rules with the possibility of procedure activation are well adapted to solve this kind of problems, where a general solution is not available but where, for several special cases algorithms are known. The system may also be used as an automatic teaching tool. A set of rules may be written to monitor the use of the system or to teach some mathematical topic.

In the case of a scientific computing system integrating symbolic, numerical and graphical capabilities a complete hierarchy integrating all kind of sets of rules described above must be constructed. This will allow to treat all interfacing problems using the same formalism.

5. Conclusion

Presently, the graphical kernel is completly implemented and is being tested. The system presents the desired characteristics of terminal independence. Control functions for four different types of terminals have been written. The task of writing one of these sets of functions does not present much difficulty and takes at most one man/day of work. On the other hand, any application runs on any type of terminal without any modification.

The choice of LISP as the base language has simplified the system implementation in many ways. Some of the advantages brought by this choice are : no need of memory management due to the use of property lists to store control data; no need to limit the number of entities like workstations, segments, transformations, etc, due to the list characteristics, and a modular organization with little interdependence among modules. The fact that the graphical system is written in LISP simplifies also the integration of this module into the rest of the system, which will be written in LISP.

The graphical applications development prove that GKS primitives are well adapted to implement this kind of interface. The concepts of segment, segment transformation, virtual input device and workstation have greatly facilitated the implementation of these interactive tools. These applications have been integrated into the REDUCE 3.2 (Hearn, 1985) system. To invoke one of the applications, from the REDUCE's algebraic mode, the user should enter an escape character and the name of the desired application. The system asks then for the parameters values required by the graphical function and executes it. When the function is terminated the control and the function's value are returned to REDUCE's algebraic mode.

The intelligent interface sketched above is now in specification phase. The first step in the implementation will be the data manipulation module which will support the working memory management. A first description of this interface formalism is given in Bittencourt (1987).

An important part of the future work is the construction of the rules, since the contents of these rules will determine the performance and the usefulness of the system. These contents must also obey to the established standards of graphical and numerical manipulations. Since no standard does exist in the algebraic algorithms area, the implementation of this system may help to establish the preliminary characteristics of such a standard for the part dealing with graphical algorithms because the approach is based upon standardized tools. Also, the intelligent interface sketched above may be instrumental to design a model for apprentice capabilities on such a system.

Acknowledgements

We would like to thank M. Jacques Calmet, Denis Lugiez and Hubert Comon for their precious advices, suggestions and comments. The referees have also provided very helpfull comments on this paper.

References

BITTENCOURT, G., *A graph formalism for knowledge representation*, to appear, 1987.

CALMET, J. and LUGIEZ, D., *A knowledge-based system for Computer Algebra*, ACM-SIGSAM Bulletin, Vol. 21, Num. 1, (Issue # 79), February 1987.

CHAR, B.W., GEDDES, K.O., GONNET, G.H. and WATT, S.M., *MAPLE user's guide*, WATCOM Publications Limited, Waterloo, Ontario, 1985.

FIKES, R. and KEHLER, T., *The role of frame-based representation in reasoning*, Communications of the ACM, volume 28, number 9, pp. 904-920, September 1985.

FODERARO, J.K. and SKLOWER, K.L., *The Franz Lisp manual*, University of California, California, April 1982.

FOLEY, J. and VAN DAN, A., *Fundamentals of interactive Computer Graphics*, Addison-Wesley, Reading, Mass., 1982.

HEARN, A.C., *REDUCE user's manual version 3.2*, The Rand Corporation, Santa Monica, April 1985.

HOPGOOD, F.R.A., DUCE, D.A., GALLOP, J.R. and SUTCLIFFE, D.C., *Introduction to the Graphical Kernel System (GKS)*, Academic Press, 1983.

KRÜGER, W., LOOS, R., PROTZEN, M. and SHULZ, T. A., *A MODULA-2 language binding for the Graphical Kernel System*, Interner Bericht Nr. 11/85, Fakultät für Informatik, Universität Karlsruhe, June 1985.

LAMPORT, L., *LATEX, user's guide & reference manual*, Addison-Wesley Publishing Company, 1986.

MARTIN, W.A., *Computer input/output of mathematical expressions*, In : "2nd Symposium on Symbolic and Algebraic Manipulation", ACM Publisher, Los Angeles, California, pp. 78-89, March 1971.

MARTIN, W.A. and FATEMAN, R.J., *The MACSYMA System*, In : "2nd Symposium on Symbolic and Algebraic Manipulation", ACM Publisher, Los Angeles, California, March 1971.

WILLIAMSON, H., *Algorithm no. 420, hidden line plotting program*, Communications of the ACM, volume 15, number 2, pp. 100-103, February 1972.

WRIGHT, T.J., *A two-space solution to the hidden-line problem for plotting functions of two variables*, IEEE Transactions C-22, number 1, pp. 28-33, January 1973.

Appendix

In this appendix we present the names and parameters of the implemented Graphical Kernel System primitives. A complete description of the functions of these primitives may be found in Hopgood et al. (1983). The primitives are grouped in modules depending on their functions. Each primitive's parameter is followed by a type name delimited by the "<" and ">" characters, this types are :

<ident> - a non numerical LISP atom.

<rectangle> - a four numerical elements list.

<pair> - a LISP dotted-pair, where the two elements are numerical.

<real> - a floating-point number.

<integer> - an integer number.

<string> - a characters string.

<points list> - a LISP A-list, where each element has numerical components.

<file> - an output port identificator.

1. Module GKS

OpenGKS (file : <ident>) : open a GKS section.

CloseGKS ( ) : close a GKS section.

InquireOperatingStateValue ( ) : return the present operating state of the system.

InquireLevelOfGKS ( ) : return the level of the present implementation.

2. Module Workstations

OpenWS (ws, comm, type : <ident>) : open a workstation of the given type and call it ws.

ActivateWS (ws : <ident>) : activate the workstation ws.

CloseWS (ws : <ident>) : close the workstation ws.

DeactivateWS (ws : <ident>) : deactivate the workstation ws.

ClearWS (ws, cm : <ident>) : send a clar display message to workstation ws.

SetDeferralState (ws, dm, igm : <ident>) : specify the deferral state of workstation ws.

UpdateWS (ws, rf : <ident>) : update the output of workstation ws.

InquireDefaultDeferralStateValues (type : <ident>) : return the default deferral state of workstations of the given type.

InquireListOfAvailableWSTypes ( ) : return the list of available types of workstations.

InquireMaximumDisplaySurfaceSize (type : <ident>) : return the maximum display surface size for workstations of a given type.

InquireSetOfOpenWS ( ) : return the list of the open workstation identificators.

InquireSetOfActiveWS ( ) : return the list of the activated workstation identificators.

InquireWSCategory (type : <ident>) : return the category of the workstation type.

InquireWSClassification (type : <ident>) : return the classification of the workstation type.

InquireWSConnectionAndType (ws : <ident>) : return the connection and type of the given workstation.

InquireWSDeferralAndUpdateStates (ws : <ident>) : return the deferral and update states of the given workstation.

InquireWSState (ws : <ident>) : return the present state of the given workstation.

3. Module Transformations

SetClippingIndicator (n : <ident>) : specify the state of the clipping indicator.

SetWindow (nt : <ident> rec : <rectangle>) : set the window size of the normalizing transformation nt.

SetViewport (nt : <ident> rec : <rectangle>) : set the viewport size of the normalizing transformation nt.

SetWSWindow (ws : <ident> rec : <rectangle>) : set the window size of the workstation ws.

SetWSViewport (ws : <ident> rec : <rectangle>) : set the viewport size of the workstation ws.

SelectNT (nt : <ident>) : select the normalizing transformation nt.

SetViewportInputPriority (n1, n2, hl : <ident>) : set a priority order for the normalizing transformations n1 and n2.

AccumulateTransMat (mi : <ident> pf, sv : <pair> ar : <real> sf : <pair> sw, mo : <ident>) : combine a transformation matrix result with the given matrix mi.

EvaluateTransMat (pf, sv : <pair> ar : <real> sf : <pair> sw, mo : <ident>) : calculate a transformation matrix.

InquireClippingIndicator ( ) : return the present state of the clipping indicator.

InquireCurrentNT ( ) : return the identificator of the present normalizing transformation.

InquireListOfNT ( ) : return the list of the defined normalizing transformations.

InquireNT (nt : <ident>) : return the window and viewport size of the given normalizing transformation.

InquireWSTransformation (ws : <ident>) : return the window and viewport size of the given workstation.

4. Module Outputs

Message (ws : <ident> st : <string>) : output a message on workstation ws.

5. Module Polylines

Polyline (pts : <points list>) : output of a line defined by the a-list pts.

SetLineType (n : <integer>) : select the line type according to the parameter n.

InquirePolylineFacilities (ws : <ident>) : return the polyline primitive characteristics.

6. Module Polymarker

Polylmarker (pts : <points list>) : output a set of marks according to the points defined by the a-list pts.

SetMarkerType (n : <integer>) : select the marker type according to the parameter n.

InquirePolylmarkerFacilities (ws : <ident>) : return the polymarker primitive characteristics.

7. Module Areas

Fillarea (pts : <points list>) : output the filled polygon defined by the a-list pts.

SetAreaStyle (n : <ident>) : select the filling style according to the parameter n.

InquireFillaresFacilities (ws : <ident>) : return the fillarea primitive characteristics.

8. Module Texts

Text (pt : <pair> st : <string>) : output a text in the given location.

SetCharacterHeight (n : <integer>) : specify the height of a character.

SetCharacterUpVector (x, y : <integer>) : specify the position for outputing characters.

SetCharacterExF (n : <integer>) : select the expansion factor for characters.

SetCharacterSpc (n : <integer>) : select the spacing between characters.

SetTextPath (p : <ident>) : select the output direction of text.

SetTextAlignment (h, v : <ident>) : select the alignement mode for text.

SetTextFP (n : <ident>) : select the font and precision for characters.

InquireTextExtent (ws : <ident> pt : <pair> st : <string>) : return the coordinates of the rectangle definig the text output.

InquireTextFacilities (ws : <ident>) : return the text primitive characteristics.

9. Module Choices

RequestChoice (ws, dv : <ident>) : return a choice primitive result from workstation ws.

10. Module Strings

RequestString (ws, dv : <ident>) : return an input stringfrom workstation ws.

11. Module Locators

RequestLocator (ws, dv : <ident>) : return a locator primitive result from workstation ws.

12. Module Pickers

RequestPick (ws, dv : <ident>) : return a picker primitive result from workstation ws.

13. Module Strokes

RequestStroke (ws, dv : <ident>) : return a list of input points from workstation ws.

14. Module Valuators

RequestValuator (ws, dv : <ident>) : return a valuator primitive result from workstation ws.

15. Module Segments

CreateSegment (ns : <ident>) : open a segment named ns.

DeleteSegment (ns : <ident>) : dlete the segment ns.

CloseSegment ( ) : close the open segment.

RenameSegment (old, new : <ident>) : change the name of a segment.

DeleteSegmentFromWS (ws, ns : <ident>) : delete the segment ns from workstation ws.

RedrawAllSegmentsOnWS (ws : <ident>) : output all segments stored in the workstation ws.

AssociateSegmentWithWS (ws, ns : <ident>) : store segment ns into workstation ws.

CopySegmentToWS (ws, ns : <ident>) : output segment ns to workstation ws without storing it as a segment.

InsertSegment (ns , nmat : <ident>) : output segment ns after transforming it according to the matrix nmat.

SetPickIdentifier (np : <ident>) : define a pick identifier inside a segment.

SetDetectability (ns, detec : <ident>) : define the detectability of the segment ns.

SetSegmentTrans (ns, matrix : <ident>) : define the tranformation matrix for segment ns.

SetVisibility (ns, vis : <ident>) : define the visibility of segment ns.

InquireNameOfOpenSegment ( ) : return the name of the open segment.

InquireSegmentAttributes (ns : <ident>) : return the attributes of segment ns.

InquireSetOfAssociatedWS (ns : <ident>) : return the list of workstations associated to the segment ns.

InquireSetOfSegmentNamesInUse ( ) : return the list of all segment identificators in use.

InquireSetOfSegmentNamesOnWS (ws : <ident>) : return the list of segment identificators associated to the workstation ws.

16. Module Errors

EmergencyCloseGKS ( ) : close all segments and workstations and quit GKS.

ErrorHandling (noe : <integer> fnt : <ident> efl : <file>) : store the error number noe detected in function fnt in the error file efl.

ErrorLogging (noe : <integer> fnt : <ident> efl : <file>) : treat the error noe detected in function fnt and output results infile efl.

Type Inference Using Unification in Computer Algebra

J. Calmet, H. Comon and D. Lugiez

LIFIA
46 Avenue Félix Viallet, 38031 Grenoble Cedex, France

(Extended abstract)

This work considers the problem of type inference in Computer Algebra systems. In a first part the reasons underlying the need for strong typing and type inference in this field are outlined. Then, the peculiarities of type inference in the framework of computer algebra systems are emphasized. Finally, a first solution to the problem is briefly sketched. A detailled description will be found in {1}. The purpose of this extended abstract is to inform of the results already obtained whithout entering into technical details.

1 - Why strong typing and type inference?

Originally, computer algebra systems consisted in a collection of algebraic algorithms allowing to perform some mathematical operations on some relevant domains. New generation systems intend to incorporate (almost) all operators to compute on (almost) all domains. This implies typing and the possibility to access the "right" definition of an operator depending on its type (problems of overloading and polymorphic operators). Also, two seemingly contradictory goals must be achieved. The first one is to allow users to input untyped expressions. The second one is that at compile time each object must have an associated type to insure correctness and efficiency. The solution is a type inference mechanism which determines the type of any expression. A first system going in this direction is Scratchpad {2}.
A new system, still in the design phase {3}, will have many more capabilities by integrating software tools, graphical and numerical computing and knowledge manipulation techniques coming from Artificial Intelligence. It will thus be much larger than present systems and consistency of the different modules becomes a prerequisite.
These remarks lead to the conclusions that strong typing is mandatory and that a type inference algorithm is much desirable in computer algebra. Then, a question arises: are the typing algorithms and methods already known, usable for our purposes?

2 - Type Inference in Computer Algebra

The most appropriate typing algorithm seems to be found in the ML language {4}, but in computer algebra the problem is more complex due to the presence of sub-types and properties. A more precise description of the type inference in the context of polymorphism, domains and domains constructors and properties has thus to be given.

Basically, in Computer Algebra, the type checking mechanism checks that the arguments of an operator have the correct types and computes a plausible type of an expression. Indeed, one deals with expressions whose values belong to domains of computation. The class of the domains contains basic domains and the domains constructed by means of domain constructors which are seen as operators. The set of types can then be modelized as an algebra of terms together with a mapping which associates to each operator a sequence of properties. Properties are classes of domains having a common mathematical structure. The properties may be used to give, as in Scratchpad, a "syntax" to the domain constructors viewed then as functors. They are also used for performing "conditional enrichments" such as : if t is a field, then we define an euclidian division on UP[t,x]. But, moreover, the properties may be used for the passage of the operators of the parameters such as in LPG {5}. This approach is different from Scratchpad II since no precision about the representation of the type has to be given. Then, it is possible to have several implementations of a same operator. Furthermore, it is possible to use the same domain constructor with the same domain parameters and distinct operator parameters through generic enrichment. Also, in the case of either polymorphic or overloaded operators, it is necessary to know the types of the arguments of the operator in order to find which operator is denoted by the symbol.

In ML {4} the type inference algorithm, called the *W* - algorithm, relies on several type inference rules and on a unification algorithm. A main result from Milner is that an expression has a most general type, which means that all the types that can be assigned to this expression are substitution (replacing types variables by types) instances of this most general type. In Computer Algebra, the type inference problem is more complex because many coercions of types are realized and have to be taken into account. The modelization of this fact leads to introduce a relation in the set of type expressions which is the subtyping relation. The type inference mechanism has thus to deal with this relation, and to realize the necessary coercions when they are needed. According to the Milner's result about type inference without subtyping, the problem of the type inference in Computer Algebra then reduces to the problem of unification modulo a relation, more precisely an ordering relation.

3 - The proposed model

In {1}, it is shown how to describe an algebraic model which suits the problem of type inference with subtyping and properties. It is then reduced to the problem of solving a system of equations in an algebra of terms. By analogy with Milner's algorithm, we could say say that we use an algorithm of "unification modulo an ordering relation". It must be noted that this is different from many-sorted unification {6}.
We give now a sketch of the results coming from what we think is the first thorough investigation of the type inference problem in Computer Algebra. We omit technical details and use only well known concepts of the domain. The types are elements of an order-sorted algebra of terms, the sorts represent the properties and the subtyping relation is almost described by a rewrite rule, R, which is the ordering relation.
A first result is to show that this type inference problem is undecidable even when properties are not considered. In that respect, it is enough to notice that the ordering relation R may be splitted into two parts. The first part contains the congruence part of R while the second part is what remains. But, it is well known that unification modulo a congruence relation is undecidable. Thus, such is the problem. The next step is to make a simplifying hypothesis and to consider a model where the congruence part of R reduces to an equality. But, even in this simplified model, undecidability cannot be lifted. This comes from the following proposition:

Proposition: Unification modulo an ordering relation which is compatible with substitutions is undecidable.

At this stage, it is obvious that within our model, it is necessary to rely on semi-decision procedures for type inference, which means that they may not terminate. Taking the properties into account again, it is possible to design two such procedures. A first complete one arises when no further assumption is made. Starting from the types to be unified, it consists in constructing the associated derivation trees in a breadth first way. Since both the proof and the resulting algorithm are very technical and rely on numerous definitions, we refer to {1} for a more precise presentation. This algorithm has a large complexity and thus a "better" one was sought for. The basic idea for this second procedure is to refine the first one for the case of linear terms (terms in which no variable appears twice). Instead of considering all the possible successors of the instances of a term in the derivation trees, one thus considers only the instances to an application of a rule. This is close to the the narrowing process of Hullot {7} except that a reduction at an occurence of a variable is allowed. In this case the resulting type inference algorithm has a much better complexity.
Both procedures are complete and compute all the common types of two given types. In practice, we need only to get one common type. This implies to perform the unification step during the construction of the trees and to stop the process as soon as a solution is found. This last procedure is of practical use.

This study gives the framework of a simplified but implementable model for type inference in computer algebra. Since the problem is not decidable in its whole generality, technical implementation details play an important role. For instance, since we have transition rules instead of true rewrite rules, it is not possible to use most of the simplifications permitted in the usual unification processes. Further studies are to be pursued in order to refine this model. Also, types for functions have not been considered since they probably imply higher order unification. A last remark is that since a simplified model has been designed there is possibly still room for improvements.

References

1 - H. Comon, D. Lugiez and Ph. Sshnoebelen, *Type Inference in Computer Algebra.* To appear in the proceedings of ISSAC '87, Leipzig, DDR; LNCS, Springer-Verlag. A long version will appear as an internal LIFIA report.

2 - A. Fortenbacher, R.D. Jenks, M. Lucks, R.S. Sutor, B.M. Trager and S.M. Watt, *An Overview of the Scratchpad II Language and System.* IBM Research Report, July 1985.

3 - J. Calmet and D. Lugiez, *A Knowledge-Based System for Computer Algebra.* ACM-SIGSAM Bulletin, **22(1)**, pp. 7-13, 1987.

4 - R. Milner, *A Theory of Type Polymorphism in Programming.* J.C.S.S. **17**, pp348-375, 1978.

5 - D. Bert and R. Echahed, *Design and Implementation of a Generic, Logic and Functional Programming Language.* Proc. of ESOP, LNCS **213**, pp 119-132, 1985.

6 - M. Schmidt-Schauss, *Unification in Many Sorted Equational Theory.* 8-th CADE Conference, LNCS **230**, pp 538-552, 1986.

7 - J.M. Hullot, *Canonical Forms and Unification.* 5th CAD, LNCS **87**, pp 318-334, 1980.

The Weight Distribution of Double-Error-Correcting Goppa Codes

Arne Dür

Institut für Mathematik, Universität Innsbruck
Techniker-Straße 25, A-6020 Innsbruck, Austria

Abstract: In this paper the weight distribution of binary Goppa codes with location set $GF(2^m)$ and irreducible quadratic Goppa polynomial is studied. These codes are all equivalent and have parameters $[2^m, 2^m-2m, 5]$. An explicit formula for the number of codewords of weight 5 and 6 is derived. The weight distribution of the dual codes is related to the weight distribution of a reversible irreducible cyclic code of length 2^m+1 and dimension 2m whose weights can be expressed in terms of Kloosterman sums over $GF(2^m)$. As numerical examples the weight distribution of the double-error-correcting Goppa codes of block length 8,16,32, and 64 are computed.

I. INTRODUCTION

For $m \geq 3$ let $\Gamma(m)$ denote the set of all binary Goppa codes with location set $GF(2^m)$ whose Goppa polynomial is an irreducible quadratic polynomial over $GF(2^m)$. It is well-known (MacWilliams and Sloane (1977)) that all codes $C \in \Gamma(m)$ are equivalent and have parameters

$$[n=2^m, k=2^m-2m, d=5]$$

(The comparable primitive BCH-codes with parameters $[2^m-1, 2^m-1-2m, 5]$ have one fewer information symbol). In particular the packing radius of these Goppa codes is e=2. Moreno (1981) proved that, when m is odd, the codes $C \in \Gamma(m)$ are quasi-perfect. Feng and Tzeng (1984) determined the covering radius (c=3 if m is odd, and c=4 if m is even), and also gave a method for the complete decoding of such Goppa codes. Berlekamp and Moreno (1973) and Tzeng and Zimmermann (1975) showed that the codes $\hat{C}$ obtained from $C \in \Gamma(m)$ by adding an overall parity check are cyclic with zeros ω^i, $i=0,\pm 1,\pm 2,\pm 4,\ldots,\pm 2^{m-1}$, ω being a primitive (2^m+1)-th root of unity in $GF(2^{2m})$.

In this article we study the weight distribution $(a(i))_{i=0}^{n}$ of $C\epsilon\Gamma(m)$. First we establish an explicit formula for $a(5)$, the number of codewords of minimum weight. Since the extended codes are cyclic, this also gives a formula for $a(6)$. Using a result of Cohen and Godlewski (1975), we obtain the series expansions of the probabilities of decoding and of erroneous decoding for an incomplete decoder on a binary symmetric channel in terms of the transition probability up to order 4.

Secondly we relate the weight distribution of the dual Goppa codes to the weight distribution of an irreducible cyclic code C_{ir} of length 2^m+1 with nonzeros ω^i, $i=\pm 1,\pm 2,\pm 4,\ldots,\pm 2^{m-1}$. Using a technique based on Gauss sums which has been introduced by McEliece and Rumsey (1972), the weight of a codeword of C_{ir} can be written as a Kloosterman sum over $GF(2^m)$. There results a quick method of computing the weight distribution of C_{ir} from the absolute trace of $GF(2^m)$, and we give the results up to $m=15$. From the MacWilliams identity we finally obtain the weight distribution of $C\epsilon\Gamma(m)$ for $m=3,4,5$ and 6.

II. THE NUMBER OF CODEWORDS OF MINIMUM WEIGHT

Let $m\geq 3$, $q=2^m$, and let $(a(i))_{i=0}^{q}$ denote the weight distribution of $C\epsilon\Gamma(m)$.

THEOREM 1: If m is odd, then

$a(5) = q(q-2)(q-3)/120$ and $a(6) = q(q-2)(q-3)(q-5)/720$.

If m is even, then

$a(5) = q(q-1)(q-4)/120$ and $a(6) = q(q-1)(q-4)(q-5)/720$.

For instance,

m :	3	4	5	6	7	8
q :	8	16	32	64	128	256
a(5):	2	24	232	2016	16800	137088
a(6):	1	44	1044	19824	344440	5734848 .

We will prove the theorem at the end of this section. First we apply a result of Cohen and Godlewski (1975), p.328, on error probabilities to obtain the following consequence.

COROLLARY: Consider a binary symmetric channel with transition probability β and an incomplete decoder for $C\in\Gamma(m)$ which uses nearest neighborhood decoding up to the packing radius e=2. Let P_d and P_{ed} denote the probabilities of decoding and of erroneous decoding, respectively. Then the series expansions of P_d and P_{ed} in terms of β up to order 4 are

$$P_d \sim 1 - \frac{1}{12}(q-2)q(q+1)\beta^3[1-\frac{3}{4}(q-3)\beta] ,$$

$$P_{ed} \sim \frac{1}{12}\, q(q-2)(q-3)\beta^3[1-\frac{3}{4}(q-3)\beta] \qquad \text{if m is odd,}$$

and

$$P_d \sim 1 - \frac{1}{12}(q-1)q^2\beta^3[1-\frac{3}{4}(q-3)\beta] ,$$

$$P_{ed} \sim \frac{1}{12}\, q(q-1)(q-4)\beta^3[1-\frac{3}{4}(q-3)\beta] \qquad \text{if m is even.}$$

For instance, when m=5, C is a [32,22,5]-code with
$P_d \sim 1 - 2640\beta^3(1- 21.75\beta)$ and $P_{ed} \sim 2320\beta^3(1- 21.75\beta)$,
whereas for the comparable [31,21,5]-BCH-code
$P_d \sim 1 - 2635\beta^3(1- 21\beta)$ and $P_{ed} \sim 1860\beta^3(1- 21\beta)$.

Proof of Theorem 1: Since the extended code $\hat{C}$ is cyclic, $a(6)=\frac{q-5}{6}a(5)$ (MacWilliams and Sloane (1977), p.232). So we only have to compute a(5). Let Ir denote the set of monic quadratic irreducible polynomials with coefficients in $GF(2^m)$. For $G\in Ir$ let C_G be the Goppa code with location set $GF(2^m)$, arranged in a fixed order, and Goppa polynomial G. Let Qu denote the set of quadruplets $\lambda=(\lambda_1,\lambda_2,\lambda_3,\lambda_4)$ with different components in $GF(2^m)$. We define a relation ρ on $Qu\times Ir$ by

$\lambda\ \rho\ G$ if and only if there is a (necessarily unique) $\gamma\in GF(2^m)$ such that $\delta(\lambda_1,\lambda_2,\lambda_3,\lambda_4,\gamma)\in C_G$.

Here $\delta(\lambda_1,\ldots,\lambda_4,\gamma)$ denotes the binary vector of length 2^m with nonzero components at locations $\lambda_1,\ldots,\lambda_4,\gamma$. Now consider the bipartite graph with vertex sets Qu,Ir and edge set ρ. The number of edges with endpoint G is (5!)a(5), and the set Ir has q(q-1)/2 elements. On the other side let $z(\lambda)$ denote the number of edges with endpoint λ. Then the total number of edges is

$$(5!)a(5)q(q-1)/2 = \Sigma_\lambda\ z(\lambda) ,$$

the summation being taken over all $\lambda\in Qu$. So it remains to determine the function $z: Qu \to \{0,1,2,\ldots\}$, $\lambda \to z(\lambda)$. The group $Aff(1,2^m)$ of affine transformations in $GF(2^m)$ operates freely on Qu by $(\lambda_i) \to (a\lambda_i+b)$, and acts transitively on Ir by $G \to a^2G((X-b)/a)$.

In order to show that the relation ρ is invariant under the induced operation of $Aff(1,2^m)$ on $Qu\times Ir$, suppose that $\lambda\ \rho\ G$. Then there exists an element $\gamma\in GF(2^m)$ such that $x=\delta(\lambda_1,\ldots,\lambda_4,\gamma)\in C_G$. By the basic theory of Goppa codes, $x\in C_G$ if and only if G divides L'_x, where L'_x denotes the derivative of the locator polynomial $L_x=(X-\lambda_1)\ldots(X-\lambda_4)(X-\gamma)$. Thus $a^2G((X-b)/a)$ divides $a^4L'_x((X-b)/a)=L'_y$ where $y=(a\lambda_1+b,\ldots,a\lambda_4+b,a\gamma+b)$, which proves the invariance of ρ. Hence $z\colon Qu\to\{0,1,\ldots\}$ is constant on the orbits of $Aff(1,2^m)$, and it suffices to compute $z(0,1,\alpha,\beta)$ for $(0,1,\alpha,\beta)\in Qu$. These values are determined in the following Lemma. Observing that X^2+X+1 is irreducible over $GF(2^m)$ if and only if m is odd, we conclude that

$$
\begin{aligned}
\Sigma_\lambda\ z(\lambda) &= (\text{order of } Aff(1,2^m))\ \Sigma_{\alpha,\beta}\ z(0,1,\alpha,\beta) = \\
&= q^2(q-1)(q-2)(q-3)/2 \qquad \text{if } m \text{ is odd,} \\
\text{or}\quad &= q^2(q-1)^2(q-4)/2 \qquad \text{if } m \text{ is even.}
\end{aligned}
$$

Q.E.D.

LEMMA 1: For $(0,1,\alpha,\beta)\in Qu$ $\quad z(0,1,\alpha,\beta)=q/2$, except if $\beta=\alpha+1$ and $\alpha^2+\alpha+1=0$ when $z(0,1,\alpha,\beta)=0$.

Proof: By definition $z(0,1,\alpha,\beta)$ is the number of polynomials $G\in Ir$ for which there exists an element $\gamma\in GF(2^m)$ such that $x=\delta(0,1,\alpha,\beta,\gamma)\in C_G$. Using the locator polynomial $L_x=X(X-1)(X-\alpha)(X-\beta)(X-\gamma)$ and its derivative $L'_x=X^4+(\alpha+\beta+\gamma+\alpha\beta+\alpha\gamma+\beta\gamma)X^2+\alpha\beta\gamma$, we have $x\in C_G$ if and only if G divides L'_x, i.e. $G^2=L'_x$. But every polynomial $G\in Ir$ has the form X^2+cX+d where $c\neq0$ and $Tr(d/c^2)=1$. Here Tr denotes the absolute trace of $GF(2^m)$. Thus $z(0,1,\alpha,\beta)$ equals the number of elements γ in $GF(2^m)$ different from $0,1,\alpha,\beta$ such that

$$(\alpha+\beta+1)\gamma+(\alpha\beta+\alpha+\beta)\neq0 \qquad\text{and}\qquad Tr\left(\frac{\alpha\beta\gamma}{(\alpha^2+\beta^2+1)\gamma^2+(\alpha^2\beta^2+\alpha^2+\beta^2)}\right)=1.$$

Apparently $z(0,1,\alpha,\beta)=0$ if $\beta=\alpha+1$ and $\alpha^2+\alpha+1=0$.
Next suppose that $\beta=\alpha+1$ and $\alpha^2+\alpha+1\neq0$. Here we must count the elements γ in $GF(2^m)$ different from $0,1,\alpha,\alpha+1$ such that $Tr(\xi\gamma)=1$, where $\xi=\dfrac{\alpha^2+\alpha}{\alpha^4+\alpha^2+1}$. But the bijection $\gamma\to\xi\gamma$ sends $0,1,\alpha,\alpha+1$ to $0,\xi,\xi\alpha,\xi\alpha+\xi$ all of which have trace 0. This is easily seen by using the formula

$$Tr\left(\frac{ab}{a^2+1}\right)=Tr\left(\frac{b^2+b}{a^2+1}\right), \quad a,b\in GF(2^m),\ a\neq1,$$

from Feng and Tzeng (1984), p.137. We conclude that in this case $z(0,1,\alpha,\beta)=q/2$.

Finally suppose that $\beta \neq \alpha+1$. Substituting δ for $\frac{1}{(\alpha+\beta+1)\gamma+(\alpha\beta+\alpha+\beta)}$, it follows that $z(0,1,\alpha,\beta)$ equals the number of elements δ in $GF(2^m)$ different from $\delta_1 = \frac{1}{\alpha\beta+\alpha+\beta}$, $\delta_2 = \frac{1}{\alpha\beta+1}$, $\delta_3 = \frac{1}{\alpha^2+\beta}$, $\delta_4 = \frac{1}{\alpha+\beta^2}$

in $GF(2^m) \cup \{\infty\}$ such that $Tr(a_2\delta^2+a_1\delta)=1$, where $a_2 = \frac{\alpha\beta(\alpha+\beta+\alpha\beta)}{\alpha+\beta+1}$ and and $a_1 = \frac{\alpha\beta}{\alpha+\beta+1}$. We now invoke a result on Weil sums (see Lidl and Niederreiter (1983)). By Corollary 5.35, p.220, the quadratic polynomial $f(\delta) = a_2\delta^2+a_1\delta$ takes values of trace 0 or 1 equally often because $a_2 \neq (a_1)^2$. Furthermore, each of the elements

$$f(\delta_1) = 0 \quad , \quad f(\delta_2) = \frac{\alpha\beta}{\alpha^2\beta^2+1} \quad , \quad f(\delta_3) = \frac{\alpha^2/\beta}{\alpha^4/\beta^2+1} \quad , \quad f(\delta_4) = \frac{\alpha/\beta^2}{\alpha^2/\beta^4+1}$$

in $GF(2^m) \cup \{\infty\}$ is either ∞ or has trace 0 by the formula of Feng and Tzeng stated above. Hence $z(0,1,\alpha,\beta) = q/2$. Q.E.D.

III. THE WEIGHT DISTRIBUTION OF THE DUAL CODES

For $C \in \Gamma(m)$ the dual code $C^\perp$ has length $q=2^m$ and dimension 2m.

THEOREM 2: The weight distribution of $C^\perp$ is $b(0) = 1$,
$b(2j-1) = (2^m+2-2j)t(2^{m-1}+1-j)$, $b(2j) = (2^m+1-2j)t(j)$, $j=1,\ldots,2^{m-1}$.
The numbers $t(1),\ldots,t(2^{m-1})$ will be specified in Theorem 4 in the next section.

Proof: By Berlekamp and Moreno (1973) the extended code $\hat{C}$ of length $n=2^m+1$ is cyclic with zeros ω^i, $i=0,\pm1,\pm2,\pm4,\ldots,\pm2^{m-1}$, ω being a primitive (2^m+1)-th root of unity in $GF(2^{2m})$. Hence the dual code $(\hat{C})^\perp$ has nonzeros ω^i, $i=0,\pm1,\pm2,\pm4,\ldots,\pm2^{m-1}$, and admits a direct sum decomposition

$$(\hat{C})^\perp = C_{ir} \oplus C_{re}$$

where $C_{re}=\{(00..0),(11..1)\}$ is the repetition code and C_{ir} is the irreducible cyclic code of length $n=2^{m+1}$ with nonzeros ω^i, $i=\pm1,\pm2,\pm4,..,\pm2^{m-1}$. On the other hand

$$(\hat{C})^\perp = em(C^\perp) \oplus C_{re}$$

where em is the embedding $GF(2)^q \to GF(2)^n$, $(x_1,..,x_q) \to (x_1,..,x_q,0)$.

Comparing both decompositions we get the linear isomorphism

$$\varphi : C_{ir} \to C^{\perp} \; , \; (c_1,..,c_n) \to (c_1 - c_n,..,c_q - c_n) \; .$$

By Theorem 4 in the next section the weight distribution of C_{ir} is $A(0)=1$, $A(2j-1)=0$, $A(2j)=nt(j)$, $j=1,...,2^{m-1}$, and $A(n)=0$. Given a codeword $c=(c_1,..,c_n) \in C_{ir}$ of weight $2j$, the weight of $\varphi(c)$ is $2j$ or $n-2j$, depending on whether $c_n = 0$ or 1. Since C_{ir} is cyclic, the number of codewords $c \in C_{ir}$ of weight $2j$ with last component $c_n=0$ is $\frac{n-2j}{n} A(2j) = (n-2j)t(j)$, whereas the number of codewords $c \in C_{ir}$ of weight $2j$ with last component $c_n=1$ is $\frac{2j}{n} A(2j) = 2jt(j)$. Combining these arguments we obtain the result. Q.E.D.

IV. SOME IRREDUCIBLE CYCLIC CODES AND KLOOSTERMAN SUMS

Let ω be a primitive (2^m+1)-th root of unity in $GF(2^{2m})$, and let C_{ir} denote the irreducible cyclic code of length $n=2^m+1$ with nonzeros ω^i, $i=\pm 1,\pm 2,\pm 4,...,\pm 2^{m-1}$. In this section we study the weights of C_{ir} by a method based on Gauss sums. Our principal reference is McEliece (1975), but see also McEliece and Rumsey (1972), Baumert and McEliece (1972), Baumert and Mykkeltveit (1973), Niederreiter (1977), and McEliece (1980). The facts about Gauss sums that we shall use can be found in the book of Lidl and Niederreiter (1983). We shortly repeat some definitions.

The canonical additive character of $GF(2^m)$ is defined by

$$\lambda(y) = (-1)^{Tr(y)} \qquad \text{for } y \in GF(2^m) \; ,$$

where Tr denotes the absolute trace of $GF(2^m)$. For $a,b \in GF(2^m)$ the sum

$$K(\lambda;a,b) = \Sigma_c \; \lambda(ac+bc^{-1})$$

ranging over all nonzero $c \in GF(2^m)$, is called a Kloosterman sum. The cases where $ab=0$ are trivial: If $a=b=0$, then $K(\lambda;a,b) = 2^m-1$, and if exactly one of a and b is 0, then $K(\lambda;a,b) = -1$. In the principal case where both a and b are nonzero, we have

$$K(\lambda;a,b) = K(\lambda;1,ab) \; .$$

Kloosterman sums over fields of characteristic 2 have been studied by Carlitz and Uchiyama (1957), and Carlitz (1969).

We now return to the irreducible cyclic code C_{ir}. Denoting the absolute trace of $GF(2^{2m})$ by Tr', we have a linear isomorphism

$$c : GF(2^{2m}) \to C_{ir} , \quad x \to (Tr'(x\omega^i))_{i=1}^{n} ,$$

(see, for instance, van Lint (1982)). In particular, the nonzero codewords divide themselves into 2^m-1 equivalence classes of n words each under the cyclic shift. The following theorem exhibits a close relationship between the weights of C_{ir} and Kloosterman sums. Let $GF(2^{2m})*$ denote the set of nonzero elements of $GF(2^{2m})$.

THEOREM 3: For $x \in GF(2^{2m})*$ the Hamming weight of $c(x)$ is

$$\|c(x)\| = \frac{1}{2} n + \frac{1}{2} K(\lambda;1,No(x)) .$$

Here No denotes the norm of $GF(2^{2m})$ over $GF(2^m)$ given by $No(x) = x^n$, $n = 2^m+1$.

Invoking results on Kloosterman sums we have the following consequences: By Carlitz (1969) $K(\lambda;1,1) = -2^{1+m/2}\cos(m\varphi)$ where the angle φ is determined by $\cos(\varphi) = -(1/8)^{1/2}$ and $\sin(\varphi) = (7/8)^{1/2}$. Hence

$$\|c(1)\| = \frac{1}{2}(2^m+1- 2^{1+m/2}\cos(m\varphi)) .$$

The recursion formula $K(0) = -2$, $K(1) = 1$, $K(m+2)+K(m+1)+2K(m) = 0$ for $K(m) = K(\lambda;1,1)$ gives

m :	1	2	3	4	5	6	7	8	9	10	11	12	13
n :	3	5	9	17	33	65	129	257	513	1025	2049	4097	8193
$K(\lambda;1,1)$:	1	3	-5	-1	11	-9	-13	31	-5	-57	67	47	-181
$\|c(1)\|$:	2	4	2	8	22	28	58	144	254	484	1058	2072	4006 .

If $a,b \in GF(2^m)$ are not both 0, then

$$|K(\lambda;a,b)| \leq 2^{1+m/2}$$

by Carlitz and Uchiyama (1957). It follows that all nonzero weights of C_{ir} lie in the interval on the real line with centre n/2 and radius $2^{1+m/2}$. In particular, the distance of C_{ir} is bounded below by

$$d(C_{ir}) \geq \frac{1}{2}(2^m-2^{2+m/2}+1) .$$

For instance, if m=13, C_{ir} has distance 4006 while the bound gives $d(C_{ir}) \geq 3916$.

Proof of Theorem 3: Let $N = 2^m-1$, and let β be a complex primitive N-th root of unity. For the purpose of calculating weights we may assume that $\omega = \psi^N$ where ψ is a primitive element of $GF(2^{2m})$. Let μ' be the multiplicative character of $GF(2^{2m})$ such that $\mu'(\psi) = \beta$, and let λ' denote the canonical additive character of $GF(2^{2m})$. We now apply a formula of McEliece (1975) for the weights in an arbitrary irreducible cyclic code (where in general $N=(2^k-1)/n$, n and k denoting the length and the dimension of the code, respectively). From formula (2.9) of McEliece (1975) it follows that for $x \in GF(2^{2m})^*$

$$\|c(x)\| = \frac{n}{2} - \frac{1}{2N}\Sigma_{i=0}^{N-1} G(\mu'^i,\lambda')\mu'^i(x^{-1}) \quad .$$

Here $G(\mu'^i,\lambda')$ is the Gauss sum $\Sigma_c \mu'^i(c)\lambda'(c)$ where c runs over all elements of $GF(2^{2m})^*$. As in other special cases studied by McEliece et alii (N=1,2,3,4; or N divides 2^e+1 for some e; or N is an odd prime for which 2 generates the quadratic residues mod N), also in our case the above expression can be greatly simplified by the theorem of Davenport and Hasse (see, e.g., Lidl and Niederreiter (1983)). Let $GF(2^m)$ be the subfield of $GF(2^{2m})$ generated by the element $\eta=\psi^n=No(\psi)$. Let μ be the multiplicative character of $GF(2^m)$ such that $\mu(\eta)=\beta$, and let λ denote the canonical additive character of $GF(2^m)$. Since the characters μ',λ' of $GF(2^{2m})$ are obtained by lifting μ,λ to $GF(2^{2m})$, we have $G(\mu'^i,\lambda')=-G(\mu^i,\lambda)^2$ by the Davenport-Hasse theorem. But $G(\mu^i,\lambda)^2=(\Sigma_c \mu^i(c)\lambda(c))^2=\Sigma_{c,d}\mu^i(cd)\lambda(c+d)=$
$=\Sigma_a \mu^i(a)\ (\Sigma_c \lambda(c+ac^{-1}))=\Sigma_a \mu^i(a)\ K(\lambda;1,a)$ where the summations are taken over $GF(2^m)^*$. Observing that $\mu'(x^{-1})=\mu^i(No(x)^{-1})$, we conclude that $-\frac{1}{N}\Sigma_{i=0}^{N-1} G(\mu'^i,\lambda')\mu'^i(x^{-1})=$
$=\Sigma_a K(\lambda;1,a)\ \frac{1}{N}\Sigma_{i=0}^{N-1}\mu^i(aNo(x)^{-1})=K(\lambda;1,No(x))$
which implies the result. Q.E.D.

LEMMA 2: For $a \in GF(2^m)^*$ let $Z(a)$ denote the number of elements $c \in GF(2^m)^*$ such that both c and ac^{-1} have absolute trace 0. Then

$$K(\lambda;1,a) = 4Z(a)-2^m+3 \; .$$

The function $Z : GF(2^m)^* \to \{0,1,2,...\}$ has the following properties:

(1) Z is invariant under the Galois group of $GF(2^m)$ over GF(2), i.e. $Z(a^2)=Z(a)$ for all $a \in GF(2^m)^*$.

(2) For every $a \in GF(2^m)^*$ the set $\{c \in GF(2^m)^*;\ Tr(c)=0=Tr(ac^{-1})\}$ is invariant under the transformation $c \to ac^{-1}$ of period 2 . Thus Z(a) is odd if and only if Tr(a)=0 .

(3) $\Sigma_a Z(a) = (2^{m-1}-1)^2$.

(4) For every $a \in GF(2^m)^*$
$(2^{m/2}-3)(2^{m/2}+1)/4 \le Z(a) \le (2^{m/2}+3)(2^{m/2}-1)/4$.

Proof: Denoting the number of elements $c \in GF(2^m)^*$ such that $Tr(c+ac^{-1})=0$ by l, we have $K(\lambda;1,a)= 2l-2^m+1$ by the definition of $K(\lambda;1,a)$. Now consider the map $\varphi: GF(2^m)^* \to GF(2^m)$, $\varphi(c)= c+ac^{-1}$. The image of φ consists of 0 and all $b \in GF(2^m)^*$ such that $Tr(a/b^2)=0$. Clearly $\varphi(c)=0$ if and only if $c= a^{1/2}$. Further, for any $b \in GF(2^m)^*$ such that $Tr(a/b^2)=0$, there are exactly two solutions $c \in GF(2^m)^*$ of the equation $c+ac^{-1}= b$. Hence $l= 1+2Z(a)$ and $K(\lambda;1,a)= 4Z(a)-2^m+3$. Properties (1) and (2) are obvious. Properties (3) and (4) are consequences of $\Sigma_a K(\lambda;1,a)= \Sigma_c \lambda(c) \Sigma_a \lambda(ac^{-1})= (\Sigma_c \lambda(c))^2= (-1)^2=1$ or $|K(\lambda;1,a)| \leq 2^{1+m/2}$ (Carlitz and Uchiyama (1957)), respectively. Q.E.D.

EXAMPLE: Let $GF(2^7)= GF(2)[\zeta]$ where $\zeta^7+\zeta+1=0$. Then the functions Tr and Z are given by

r :	0	1	3	5	7	9	11	13	15	19	21	23	27	29	31	43	47	55	63
Tr :	1	0	0	0	1	0	0	1	0	1	1	0	1	0	1	1	1	0	1
Z :	28	31	29	35	28	29	35	36	29	32	34	33	32	31	34	26	32	27	30 .

Here r runs through the representatives of the cyclotomic cosets mod 127. For instance, $Tr(\zeta^{12})= Tr(\zeta^3)= 0$ and $Z(\zeta^{12})= Z(\zeta^3)= 29$.

The following theorem relates the weight distribution of the irreducible cyclic code C_{ir} to the distribution of the function $Z : GF(2^m)^* \to \{0,1,2,...\}$.

THEOREM 4: For $j=1,...,2^{m-1}$ let $t(j)$ denote the number of elements $a \in GF(2^m)^*$ such that $Z(a)= j-1$. Clearly $\Sigma\, t(j) = 2^m-1$, and $\Sigma\, jt(j) = 2^{2m-2}$ by the above lemma. Then the weight distribution of C_{ir} is

$A(0)= 1$, $A(2j-1)= 0$, $A(2j)= (2^m+1)t(j)$, $j=1,...,2^{m-1}$, $A(n)= 0$.

In particular the number of equivalence classes of codewords of weight 2j under the cyclic shift is $t(j)$.

Proof: Combining Theorem 3 and Lemma 2 we have $\|c(x)\| = 2\{Z(No(x))+1\}$ for $x \in GF(2^{2m})^*$. Since for each $a \in GF(2^m)^*$ there are exactly 2^m+1 elements $x \in GF(2^{2m})^*$ such that $No(x)=a$, the result follows. Q.E.D.

Finally, we use our results on the weight distribution of the codes $C \in \Gamma(m)$ to prove a result of mathematical interest.

THEOREM 5: Let $q=2^m$. Then the number of nonzero solutions of the equations

$$x_1+x_2+x_3+x_4+x_5=1 \quad \text{and} \quad \frac{1}{x_1}+\frac{1}{x_2}+\frac{1}{x_3}+\frac{1}{x_4}+\frac{1}{x_5}=1$$

in GF(q) is $q^3+6q-19$.

Proof: The number of solutions is $\Sigma_{a,b} K(\lambda;a,b)^6/(q-1)q^2$ where the summation is taken over all elements a,b in GF(q). This can be seen by multiplying out $K(\lambda;a,b)^6$ (compare Carlitz (1969), pp.263). As $\Sigma_{a,b} K(\lambda;a,b)^6 = (q-1)^6+2(q-1)+(q-1)\Sigma_{b\neq 0} K(\lambda;1,b)^6$, it remains to show that

$$\Sigma_{b\neq 0} K(\lambda;1,b)^6 = 5q^4-4q^3-9q^2-5q-1 .$$

To this end, choose $C\in\Gamma(m)$ and let $(\hat{b}(i))_{i=0}^n$ denote the weight distribution of $(\hat{C})^\perp$. Since $(\hat{C})^\perp = C_{ir}\oplus C_{re}$ where C_{re} is the repetition code, we have $\hat{b}(0)=1=\hat{b}(n)$ and, for $j=1,2,\ldots,q/2$, $\hat{b}(2j)=A(2j)$ and $\hat{b}(2j-1)=A(n-2j+1)$. From Theorem 3 it follows that, for $i=1,2,\ldots,q$, $A(i)/n$ is the number of nonzero elements b in GF(q) such that $K(\lambda;1,b)=2i-n$. Hence $\hat{b}(i)/n$ is the number of nonzero elements b in GF(q) such that $K(\lambda;1,b)=(-1)^i(2i-n)$, which implies that $\Sigma_{b\neq 0} K(\lambda;1,b)^6 = \frac{1}{2n}\Sigma_{i=1}^q (n-2i)^6\,\hat{b}(i)$.

On the other hand, let $(\hat{a}(i))_{i=0}^n$ denote the weight distribution of $\hat{C}$. Clearly $\hat{a}(0)=1$, $\hat{a}(i)=0$ for $i=1,2,\ldots,5$, and, by Theorem 1, $\hat{a}(6)=(q+1)q(q-2)(q-3)/6!$. Then formula (24) in MacWilliams and Sloane (1977), p.132, gives

$$\Sigma_{i=0}^n (n-2i)^6\,\hat{b}(i) = 2q^2\,\Sigma_{j=0}^6 F_6^{(j)}(n)\hat{a}(j)$$

where $F_6^{(j)}(n)=d^6/dx^6[\cosh^{n-j}(x)\sinh^j(x)]_{x=0}$. Using that $F_6^{(0)}(n)=n(15q^2+1)$ and $F_6^{(6)}(n)=6!$, we conclude that $\Sigma_{b\neq 0} K(\lambda;1,b)^6 = 5q^4-4q^3-9q^2-5q-1$. Q.E.D.

V. TWO TABLES

The numbers $t(j)$, $j=1,2,\ldots,2^{m-1}$, can be calculated as follows: Choose a primitive binary polynomial P(X) of degree m, and consider the representation of $GF(2^m)$ as residue class ring of binary polynomials mod P(X). Let ζ denote the residue class of X which is a primitive element of $GF(2^m)$. Then the computation of $Tr(\zeta^r)$, r running through the representatives of the cyclotomic cosets mod 2^m-1, can be done by reducing binary polynomials mod $P(X)$. Finally a simple count gives

$$Z(\zeta^r) = \text{number of residue classes } s \bmod 2^m-1 \text{ such that } Tr(\zeta^s) = 0 = Tr(\zeta^{r-s}) ,$$

and

$$t(j) = \text{number of residue classes } s \bmod 2^m-1 \text{ such that } Z(\zeta^s) = j-1 .$$

Table 1 lists the nonzero numbers $t(j)$ for $m=1,2,\ldots,15$ in the format "$j:t(j)$" . Since for $m \leq 13$ the pairs $2j:t(j)$ can be found in the table of MacWilliams and Seery (1981), we only give the first three and the last three entries $j:t(j)$ for $m=8,9,\ldots,13$. The procedure of MacWilliams and Seery to determine the weight distribution of a nondegenerate irreducible cyclic code works by rearranging a word of the simplex code of length $2^{2m}-1$ and requires a powerful computer. For the special class of reversible irreducible cyclic codes considered here, however, the method based on the absolute trace of $GF(2^m)$ seems to be better in time and storage requirements when m becomes large.

By Theorem 2 and the MacWilliams identity we obtain the weight distribution of the codes $C \in \Gamma(m)$ of length $n=2^m$ and dimension $k=2^m-2m$. Table 2 lists the weight distribution $(a(i))_{i=0}^{n}$ of $C \in \Gamma(m)$ for $m=3,4,5,6$ in the format "$i:a(i)$" . For $m=4$ the [16,8,5]-codes $C \in \Gamma(4)$ are formally self-dual, i.e. the weight distributions of C and $C^{\perp}$ coincide. For $m=6$ the [64,52,5]-codes $C \in \Gamma(6)$ contain each about 4.5×10^{15} codewords.

TABLE 1 : Weight distribution of C_{ir} (see text for key)

m	n	j:t(j)
1	3	1:1
2	5	1:2 2:1
3	9	1:1 2:3 3:3
4	17	3:4 4:5 5:4 6:2
5	33	6:5 7:5 8:5 9:10 10:5 11:1
6	65	13:6 14:7 15:12 16:12 17:6 18:9 19:8 20:3
7	129	27:7 28:7 29:8 30:21 31:7 32:14 33:21 34:7 35:14 36:14 37:7
8	257	57:8 58:16 59:8 ... 70:20 71:8 72:5
9	513	117:3 118:9 119:18 ... 137:9 138:18 139:9

m	n	j:t(j)
10	1025	241:12 242:11 243:30 ... 270:30 271:20 272:5
11	2049	490:11 491:11 492:44 ... 532:33 533:11 534:22
12	4097	993:24 994:24 995:28 ... 1054:30 1055:12 1056:12
13	8193	2003:1 2004:39 2005:26 ... 2091:65 2092:13 2093:13
14	16385	4033:14 4034:28 4035:56 4036:56 4037:28 4038:56 4039:112 4040:84 4041:140 4042:56 4043:56 4044:56 4045:140 4046:84 4047:84 4048:168 4049:56 4050:189 4051:140 4052:84 4053:112 4054:140 4055:84 4056:112 4057:140 4058:84 4059:238 4060:196 4061:84 4062:112 4063:84 4064:112 4065:140 4066:224 4067:126 4068:140 4069:196 4070:133 4071:112 4072:84 4073:98 4074:224 4075:252 4076:140 4077:126 4078:112 4079:140 4080:280 4081:224 4082:84 4083:112 4084:140 4085:168 4086:224 4087:168 4088:119 4089:112 4090:168 4091:140 4092:168 4093:182 4094:112 4095:350 4096:112 4097:112 4098:112 4099:196 4100:140 4101:196 4102:252 4103:70 4104:294 4105:168 4106:112 4107:112 4108:168 4109:112 4110:168 4111:140 4112:84 4113:224 4114:168 4115:126 4116:168 4117:140 4118:120 4119:168 4120:224 4121:112 4122:140 4123:154 4124:84 4125:224 4126:168 4127:84 4128:112 4129:252 4130:154 4131:168 4132:112 4133:84 4134:112 4135:140 4136:140 4137:84 4138:140 4139:112 4140:189 4141:112 4142:56 4143:112 4144:112 4145:84 4146:112 4147:112 4148:63 4149:70 4150:140 4151:98 4152:56 4153:56 4154:49 4155:84 4156:84 4157:42 4158:63 4159:30 4160:14
15	32769	8102:15 8103:45 8104:30 8105:60 8106:90 8107:90 8108:60 8109:75 8110:95 8111:45 8112:225 8113:60 8114:60 8115:285 8116:75 8117:105 8118:165 8119:120 8120:150 8121:180 8122:123 8123:90 8124:165 8125:105 8126:120 8127:250 8128:135 8129:150 8130:255 8131:90 8132:180 8133:255 8134:105 8135:120 8136:255 8137:165 8138:105 8139:195 8140:210 8141:180 8142:390 8143:150 8144:90 8145:330 8146:105 8147:150 8148:300 8149:120 8150:240 8151:240 8152:150 8153:90 8154:360 8155:180 8156:120 8157:375 8158:135 8159:135 8160:330 8161:180 8162:300 8163:300 8164:150 8165:225 8166:195 8167:165 8168:120 8169:210 8170:180 8171:150 8172:360 8173:120 8174:105 8175:555 8176:150 8177:270 8178:300 8179:120 8180:180 8181:300 8182:180 8183:135 8184:360 8185:300 8186:105 8187:345 8188:135 8189:150 8190:360 8191:120 8192:285 8193:225 8194:195 8195:180 8196:420 8197:240 8198:180 8199:285 8200:150 8201:120 8202:270 8203:195 8204:120 8205:510 8206:165 8207:225 8208:225 8209:135 8210:195 8211:300 8212:180 8213:165 8214:315 8215:180 8216:155 8217:525 8218:150 8219:120 8220:525 8221:90 8222:165 8223:213 8224:180 8225:210 8226:245 8227:285 8228:105 8229:240 8230:165 8231:150 8232:420 8233:150 8234:135 8235:360 8236:105 8237:150 8238:315 8239:120 8240:210 8241:165 8242:180 8243:90 8244:135 8245:180 8246:120 8247:360 8248:105 8249:120 8250:375 8251:90 8252:195 8253:270 8254:90 8255:195 8256:195 8257:90 8258:90 8259:210 8260:120 8261:91 8262:375 8263:90 8264:90 8265:240 8266:90 8267:90 8268:120 8269:90 8270:75 8271:165 8272:105 8273:90 8274:90 8275:105 8276:60 8277:120 8278:30 8279:45 8280:120 8281:35 8282:30.

TABLE 2 : Weight distribution of $C \in \Gamma(m)$

m	n	k	i:a(i)
3	8	2	0:1 5:2 6:1
4	16	8	0:1 5:24 6:44 7:40 8:45 9:40 10:28 11:24 12:10 (formally self-dual)
5	32	22	0:1 5:232 6:1044 7:3200 8:10000 9:27410 10:63043 11:126348 12:221109 13:338450 14:459325 15:553104 16:587673 17:552708 18:460590 19:338120 20:219778 21:127228 22:63614 23:26640 24:9990 25:3562 26:959 27:140 28:25 29:10 30:1
6	64	52	0:1 5:2016 6:19824 7:150816 8:1 074564 9:6 724292 10:36 983606 11:181 562148 12:801 899487 13:3207 100848 14:11683 010232 15:38945 583600 16:119270 849775 17:336759 114600 18:879315 465900 19:2 128877 967720 20:4 789975 427370 21:10 036132 584720 22:19 616077 324680 23:35 820653 775120 24:61 193616 865830 25:97 909825 691292 26:146 864738 536938 27:206 698450 488572 28:273 137238 145613 29:339 067000 563552 30:395 578167 324144 31:433 859858 044896 32:447 417978 608799 33:433 859884 789296 34:395 578130 249064 35:339 066984 516912 36:273 137293 083068 37:206 698451 825792 38:146 864689 455168 39:97 909835 209152 40:61 193647 005720 41:35 820640 796220 42:19 616065 197930 43:10 036143 286620 44:4 789977 477705 45:2 128871 501040 46:879316 489560 47:336762 120240 48:119269 917585 49:38944 498920 50:11683 349676 51:3207 401448 52:801 850362 53:181 500048 54:36 972232 55:6 733328 56:1 082142 57:149988 58:18102 59:2052 60:171 (large integers are given in groups of six digits).

REFERENCES

BAUMERT, L.D., and MCELIECE, R.J. (1972), Weights of irreducible cyclic codes, Inform. and Control 20, pp.158-175.

BAUMERT, L.D., and MYKKELTVEIT, J. (1973), Weight distributions of some irreducible cyclic codes, DSN Progress Report 16, pp.128-131 (published by Jet Propulsion Laboratory, Pasadena, California).

BERLEKAMP, E.R., and MORENO, O. (1973), Extended double-error-correcting binary Goppa codes are cyclic, IEEE Trans. Info. Theory 19, pp.817-818.

CARLITZ, L. (1969), Gauss sums over finite fields of order 2^n, Acta Arithmetica 15, pp.247-265.

CARLITZ, L. (1969), Kloosterman sums and finite field extensions, Acta Arithmetica 16, pp.179-193.

CARLITZ, L., and UCHIYAMA, S. (1957), Bounds for exponential sums, Duke Math. J. 24, pp.37-41.

COHEN, G.D., and GODLEWSKI, P.J. (1975), Residual error rate of binary linear block codes, in: CISM Lecture Notes 219 (G. Longo, ed.), pp.325-335, Springer, Wien-New York.

FENG, G.L., and TZENG, K.K. (1984), On quasi-perfect property of double-error-correcting Goppa codes and their complete decoding, Inform. and Control 61, pp.132-146.

LIDL, R., and NIEDERREITER, H. (1983), Finite Fields, Addison-Wesley, Reading, Massachusetts.

MACWILLIAMS, F.J., and SEERY, J. (1981), The weight distributions of some minimal cyclic codes, IEEE Trans. Info. Theory 27, pp.796-806.

MACWILLIAMS, F.J., and SLOANE, N.J.A. (1977), The Theory of Error-Correcting Codes, North Holland, Amsterdam.

MCELIECE, R.J. (1975), Irreducible cyclic codes and Gauss sums, in: Combinatorics (M. Hall, Jr., and J.H. van Lint, eds.), pp.185-202, Reidel, Dordrecht-Boston.

MCELIECE, R.J. (1980), Correlation properties of sets of sequences derived from irreducible cyclic codes, Inform. and Control 45, pp.18-25.

MCELIECE, R.J., and Rumsey, H. (1972), Euler products, cyclotomy, and coding, J. Number Theory 4, pp.302-311.

MORENO, O. (1981), Goppa codes related quasi-perfect double-error-correcting codes, in "Abstracts of Papers", IEEE Internat. Sympos. Inform. Theory, Santa Monica, California.

NIEDERREITER, H. (1977), Weights of cyclic codes, Inform. and Control 34, pp.130-140.

TZENG, K.K., and ZIMMERMANN, K. (1975), On extending Goppa codes to cyclic codes, IEEE Trans. Info. Theory 21, pp.712-716.

VAN LINT, J.H. (1982), Introduction to Coding Theory, Springer, New York-Heidelberg-Berlin.

A SIMPLE ANALYSIS OF THE BLOKH-ZYABLOV DECODING ALGORITHM

Thomas Ericson
Linköping University
Dept. of Electrical Engineering
S-581 83 LINKÖPING
Sweden

ABSTRACT

Blokh-Zyablov [1] devised a decoding algorithm for concatenated codes, which is capable of maximum random error correction. The algorithm was further developed by Zinoviev-Zyablov [2], [3], who modified it so that it could also correct many bursts of errors, without sacrificing the random error correcting capability. Unfortunately hitherto available analyses of the algorithm are rather involved - a fact which might have prevented the algorithm from achieving the attention it deserves. We offer here a much simplified treatment, which we hope will help to popularize the algorithm. It should be pointed out that the basic ideas can be traced back to Forney [4], [5] (generalized minimum distance decoding).

1. INTRODUCTION

In 1966 Forney published some very interesting results on a decoding principle which he called "generalized minimum distance decoding", [4], [5]. The algorithm he devised was perhaps mainly intended for soft decision decoding, but the basic ideas are of quite a general nature and thus applicable in many other contexts as well. In particular they are extremely useful in decoding of concatenated codes (another invention of Forneys [5]).

The application of generalized minimum distance decoding with respect to concatenated codes was further developed by Blokh-Zyablov [1], who modified the algorithm substantially and performed a thorough investigation of both principal and practical matters. Further important developments were performed by Zinoviev - Zyablov [2], [3], who in particular investigated the problem of combined correction of random errors and bursts of errors.

Somewhat related to this work are also the investigations by Reddy - Robinson [10] and by Weldon [11] regarding decoding of product codes. The same ideas were also used by Justesen [12].

Unfortunately all of these investigations are somewhat involved as far as the analysis is concerned. We suggest here a much simpler approach, where the decoding power of the algorithm is proved in just a few lines. Our approach is mainly based on the papers by Zinoviev-Zyablov, [2], [3]. A similar argument has also been worked out by Van Lint [13].

2. BASIC CONCEPTS

A concatenated code is a combination of two codes such that the codewords of one code (the inner code) are associated with the letters of the other (the outer code). Formally, let the outer encoding be characterized by a mapping $f\colon \mathcal{M} \to \mathcal{X}^a$, where $\mathcal{M}$ is a message set and $\mathcal{X}$ is the alphabet of the outer code. $\mathcal{X}^a$ denotes the a-fold Cartesian product; f maps messages $m \in \mathcal{M}$ into a-sequences $f(m) = x = (x_1, x_2, \ldots, x_a)^T \in \mathcal{X}^a$, which for convenience of later notation we represent as column vectors ("T" denotes transpose). The alphabet $\mathcal{X}$ of the outer code is also the message set of the inner code $g\colon \mathcal{X} \to \mathcal{U}^b$, for which $\mathcal{U}$ is the alphabet. The a-fold Cartesian extension $g^a\colon \mathcal{X}^a \to \mathcal{U}^{ab}$ maps a-sequences $x = (x_1, x_2, \ldots, x_a)^T$ into axb-matrices $U = \{u_{ij}\}$; $i = 1,2,\ldots, a$; $j = 1,2,\ldots, b$. The concatenation of the outer encoder $f\colon \mathcal{M} \to \mathcal{X}^a$ and the inner encoder $g\colon \mathcal{X} \to \mathcal{U}^b$ is defined by the mapping $F \triangleq g^a f\colon \mathcal{M} \to \mathcal{U}^{ab}$, which maps messages $m \in \mathcal{M}$ into axb-matrices $U \in \mathcal{U}^{ab}$. The idea is illustrated in fig. 1.

We characterize codes by their minimum distance (minimal Hamming distance between any two distinct codewords in the code). It is easy to see that if the outer code "f" has minimum distance d_a and the inner code "g" has minimum distance d_b, then the concatenation $F = g^a f$ has minimum distance $d \geq d_a d_b$. However, designing a decoder which employs the maximum possible error-correcting capability (i.e. one which corrects all error patterns of weight less than $\frac{1}{2} d_a d_b$) is a nontrivial task. For instance, a direct combination of a $\left\lfloor \frac{d_b - 1}{2} \right\rfloor$ - error-correcting inner decoder and a $\left\lfloor \frac{d_a - 1}{2} \right\rfloor$ error-correcting outer decoder corrects only about $\frac{d_a d_b}{4}$ errors. To be more precise: there is (as can easily be seen) at least one error pattern with weight $\left(\left\lfloor \frac{d_a - 1}{2} \right\rfloor + 1 \right) \left(\left\lfloor \frac{d_b - 1}{2} \right\rfloor + 1 \right)$ which is not corrected. The error-correcting capability of the Blokh-Zyablov decoding algorithm is maximal, which means that it corrects all error patterns of about twice that weight. We proceed to describe the algorithm.

3. DESCRIPTION OF THE ALGORITHM

The algorithm depends on two things: error-erasure decoding and the use of several branches with different tentative decisions. The general form is indicated in fig. 2. Altogether there are $\left\lfloor \frac{d_b-1}{2} \right\rfloor$ branches, each having a decoder γ_r of the form γ_r: $\mathcal{U}^b \rightarrow \tilde{X} \triangleq X \cup \{\text{erasure}\}$ as a first step. The extra symbol "erasure" in the enlarged alphabet $\tilde{X}$ provides the possibility for the decoder to abstain from making a decision. The decoder γ_r is defined as follows:

$$\gamma_r(y^b) = \begin{cases} x \in X \text{ if } d_H[y^b, g(x)] \leq r \\ \text{"erasure" otherwise} \end{cases}$$

$$r = 0,1,\ldots, \left\lfloor \frac{d_b-1}{2} \right\rfloor ,$$

where d_H denotes Hamming distance and $y^b = y_1 y_2 \ldots y_b$ is a b-sequence of symbols $y_i \in \mathcal{U}$. The description of the next step requires some additional definitions.

Suppose the codeword $U = F(m)$ is transmitted over some channel. Recall that we represent U as an axb-matrix over the alphabet $\mathcal{U}$; $U = \{u_{ij}\}$; $i = 1,2,\ldots, a$; $j = 1,2,\ldots, b$. We assume that the received symbols y are also elements of the same alphabet $\mathcal{U}$, and we represent them in a matrix $Y = \{y_{ij}\}$; $i = 1,2,\ldots, a$; $j = 1,2,\ldots,b$. The <u>error matrix</u> $e = \{e_{ij}\}$ is defined by

$$e_{ij} = \begin{cases} 1 & \text{if } Y_{ij} \neq U_{ij} \\ 0 & \text{if } Y_{ij} = U_{ij} \end{cases} .$$

Let y_i denote the i-th row in Y, i.e. let Y be of the form

$$Y = \begin{bmatrix} - y_1 - \\ - y_2 - \\ \\ - y_a - \end{bmatrix} .$$

Each vector $y_i = (y_{i1}, y_{i2}, \cdots, y_{ib})$ is of course a b-sequence. The b-sequences are processed successively by the inner decoders γ_r as described above. The resulting sequences $\tilde{x}(r) = [\tilde{x}_1(r), \tilde{x}_2(r), \ldots, \tilde{x}_a(r)]^T$ are fed to an error-erasure decoder $\varphi\colon \tilde{X}^a \to M$, which is the same for all $\left\lfloor \frac{d_b - 1}{2} \right\rfloor$ branches. This decoder - the outer decoder - is defined as follows.

For any pair of a-vectors $\tilde{x}^a \in \tilde{X}^a$; $x^a \in X^a$ let $t = t(\tilde{x}^a, x^a)$ denote the number of components "ν" for which $\tilde{x}_\nu$ is an element of X different from x_ν, and let $\rho(\tilde{x}^a)$ denote the number of components "ν" where $\tilde{x}_\nu$ is <u>not</u> an element of X, i.e. where $\tilde{x}_\nu$ is an erasure. Thus, if $\tilde{x}^a$ is regarded as an estimate for x^a, then t denotes the number of errors and ρ denotes the number of erasures in this estimate. It is well known (and easy to prove) that for any given $\tilde{x}^a \in \tilde{X}^a$ there is at most one codeword $f(m) \in X^a$ such that

$$2t(\tilde{x}^a, f(m)) + \rho(\tilde{x}^a) < d_a .$$

Conversely, any combination of t errors and ρ erasures can be corrected as long as $2t + \rho < d_a$. We define the <u>outer decoder</u> $\varphi\colon \tilde{X}^a \to M$ such that $\varphi(\tilde{x}^a) = m$ whenever an $m \in M$ satisfying this criterion exists. When none exists the definition of φ is arbitrary.

Now in order finally to define the overall decoder $\Phi\colon U^{ab} \to M$ for the concatenated code $F = g^a f$, let us first introduce γ_r^a as a notation for the a-fold Cartesian extension of γ_r. Clearly γ_r^a is a mapping of the form $\gamma_r^a\colon U^{ab} \to \tilde{X}^a$, mapping the received matrix $Y = \{y_{ij}\}$ into an a-sequence

$$\tilde{x}^a(r) = [\tilde{x}_1(r), \tilde{x}_2(r), \ldots, \tilde{x}_a(r)]^T ,$$

which, when fed to the outer decoder $\varphi: \tilde{X} \to M$, produces the tentative decision

$$\hat{m}_r \triangleq \varphi[\gamma_r^a(Y)] \ ;$$

$$r = 0,1,\ldots , \left\lfloor \frac{d_b-1}{2} \right\rfloor .$$

The final decision is performed with the aid of a metric D, which we define as follows. Denote by w_i the Hamming weight of the i-th row $e_i = (e_{i1}, e_{i2}, \ldots , e_{ib})$ of the error matrix $e = \{e_{ij}\}$, where e is related to the matrices U and Y as in the previous section. We define

$$D(U,Y) \triangleq \Gamma(e) \triangleq \sum_{i=1}^{a} \min\{w_i, d_b\} .$$

Notice that D(U,Y) is a metric (in the mathematical sense of the word, i.e. D(U,Y) is positive definite, symmetric, and satisfies the triangle inequality). Also notice that D is dominated by the Hamming metric:

$$D(U,Y) \leq d_H(U,Y) \ ; \quad (U,Y) \in \mathcal{U}^{2ab} .$$

The final decision $\hat{m} = \Phi(Y)$ is obtained by a minimization over r of the metric $D(Y,F(\hat{m}_r))$ i.e. m is chosen as one of those (there are usually more than one) tentative decisions $\hat{m}_r$ for which the corresponding codeword $F(\hat{m}_r)$ is closest to the received matrix Y in the sense of the metric D. Next let us see what performance can be achieved with this decoder.

4. DECODING CAPABILITY

Theorem:

The decoder $\Phi: \mathcal{U}^{ab} \to \mathcal{M}$ (as defined in section II) corrects any error e such that $\Gamma(e) < \frac{d_a d_b}{2}$. Formally:

$$\Gamma(e) < \frac{d_a d_b}{2} \implies \Phi(Y) = m .$$

Proof:

For any pair (ξ,η) of non-negative real numbers define

$$E(\xi,\eta) = \begin{cases} 0 & 0 \le \xi \le \eta \\ 1 & \eta < \xi < d_b - \eta \\ 2 & \xi \ge d_b - \eta \end{cases}$$

and notice the following two simple relations:

i) $\sum_{i=1}^{a} E(w_i,r) \ge 2t_r + \rho_r \; ; \quad r = 0, 1, \ldots , \left\lfloor \frac{d_b - 1}{2} \right\rfloor ;$

where $t_r(\rho_r)$ denotes the number of errors (erasures) in the r-th branch;

ii) $\int_0^{d_b/2} E(w,\eta)\, d\eta = \min\{w,d_b\} \; ; \quad w = 0,1,\ldots, b .$

The first relation is illustrated in fig. 3 while fig. 4 provides an explanation of relation ii). Utilizing these two relations we obtain:

$$\Gamma(e) \triangleq \sum_{i=1}^{a} \min\{w_i,d_b\} = \sum_{i=1}^{a} \int_0^{d_b/2} E(w_i,\eta)d\eta$$

$$= \int_0^{d_b/2} \sum_{i=1}^{a} E(w_i,\eta)d\eta \ge \frac{d_b}{2} \min_r\{2t_r+\rho_r\} .$$

Thus, whenever $\Gamma(e)$ is less than $\frac{d_a d_b}{2}$ the inequality $2t_r+\rho_r < d_a$ must hold for at least one branch r, with the consequence that the corresponding tentative decision $\hat{m}_r$ equals m. Clearly this is enough to ensure that also the final decision $\hat{m}$ is correct.

□

In order to illuminate this result let us first consider correction of random errors. As the metric D(U,Y) is dominated by the Hamming metric $d_H(U,Y)$ it follows that all errors e with Hamming weight $w_H(e) < \frac{d_a d_b}{2}$ are corrected. Thus the algorithm performs maximal random error correction. However, in addition many error patterns with weight exceeding the number $\frac{d_a d_b}{2}$ are corrected as well. For instance, suppose all errors are located in $\left\lfloor \frac{d_a-1}{2} \right\rfloor$ different rows of the received matrix $Y = \{y_{ij}\}$. We see that in such a case correction is always obtained regardless of the number of errors in those particular rows. Thus some (but of course not all!) error patterns with weight $b \cdot \left\lfloor \frac{d_a-1}{2} \right\rfloor$ are corrected. Notice that b (the length of the inner code) might be much larger than d_b (the minimum distance of the inner code).

In the general case, with both bursts and random errors, let us say that a row is bursty whenever $\left\lfloor \frac{d_b+1}{2} \right\rfloor$ or more errors occur in this row, and denote by "x" the number of bursty rows. The total number of remaining random errors is denoted by "t" (see fig. 5). From the theorem, it follows that correct decoding will take place as long as the inequality

$$2 \cdot (t+xd_b) < d_a d_b$$

is fulfilled. Again, as b might be much larger than d_b we see that the burst correcting capability obtained on top of the random error correcting capability might be quite substantial. Another interesting fact is that the above results open the way for a convenient and powerful adaptation of the code to the error statistics of the channel.

5. REMARKS ON IMPLEMENTATION

A practical implementation of the Blokh-Zyablov algorithm would probably not directly employ the structure described above, which was convenient for describing and analysing the algorithm. A moment's reflexion reveals that exactly the same operation is obtained if we implement only the lowest branch in fig. 2 (the branch with number $r = \left\lfloor \frac{d_b-1}{2} \right\rfloor$), with the modification that for each inner decoding we store also the distance $d_H(y^b,\tilde{x})$ between the input vector $y^b = y_1y_2 \dots y_b$ and the corresponding output $\tilde{x} = \gamma(y^b)$; an erasure is stored just as such. After the outer decoding the result $\hat{m} = \varphi[\gamma^a(Y)]$ is compared with the channel output Y. If $D(Y,F(\hat{m}))$ is less than $\frac{d_b d_b}{2}$ we accept $\hat{m}$ right away; if not we try another outer decoding, replacing now with erasure the least reliable ones of the decoded outputs from the inner decoder i.e. all those, if any, at distance $\left\lfloor \frac{d_b-1}{2} \right\rfloor$ from an inner codeword. If this second try also fails to produce an estimate $\hat{m}$ within D-distance less than $\frac{d_a d_b}{2}$ from the received matrix Y, we erase also the second least reliable ones of the decoded symbols from the inner decoder. If after all possible $\left\lfloor \frac{d_b+1}{2} \right\rfloor$ steps still no acceptable decision has been obtained, we might choose any of the rejected ones or announce an erasure.

The saving in complexity using this modified form of the algorithm is obvious. The resulting total complexity depends of course on the complexity of the inner and the outer decoders. Typically the inner code might be a BCH or Goppa code for which efficient decoding algorithms are known, (we refer in particular to Blahut [6]). The outer code is most naturally chosen as a Reed-Solomon code, and again efficient decoding can be implemented (Blahut [6], see also ref. [7], [8]).

The strength of code concatenation is that it admits construction and implementation of very long codes (say of length 10^3 or more) with a reasonable amount of complexity. The point is that the complexity is determined by the lengths "a" and "b" of the outer and inner codes individually rather than by the total length n = ab. The question of complexity of construction and implementation of concatenated codes is discussed extensively in chapter 6 of the book by Blokh-Zyablov [9].

6. ACKNOWLEDGEMENT

This article is dedicated to Victor Zyablov and Victor Zinoviev who - together with the late professor Blokh - originally derived all the results of this paper, and who taught the author about them during two different stays in Linköping (Zyablov in 1984, Zinoviev in 1985). Thanks are also due to Mr. Suhail Fawakhiri, who produced an unofficial translation to English of the book by Blokh - Zyablov [9] during his period as a graduate student at the Institute of Technology in Linköping.

REFERENCES

[1] E.L. Blokh and V.V. Zyablov, "Generalized concatenated codes", (in Russian), Svyaz', Moscow, 1976.

[2] V.A. Zinoviev and V.V. Zyablov, "Decoding of non-linear generalized concatenated codes", Probl. Peredachi Inf., Vol. 14, No. 2, pp. 46-52, 1978.

[3] V.A. Zinoviev and V.V. Zyablov, "Correction of error bursts and independent errors by generalized concatenated codes", Probl. Peredachi Inf., Vol. 15, No. 2, pp. 58-70, 1979.

[4] G.D. Forney Jr., "Generalized minimum distane decoding", IEEE Trans. Inf. Theory, IT-12, pp. 125-131, 1966.

[5] G.D. Forney Jr., "Concatenated codes", MIT Research Monograph No. 37, The MIT Press, Cambridge, Mass. 1966.

[6] R.E. Blahut, "Theory and practice of error control codes", Addison Wesley 1983.

[7] I. Hsu, I.S. Reed, T.K. Truong, K. Wang, C. Yeh, and L.J. Deutsch, "The VLSI-implementation of a Reed-Solomon encoder using Berlekamps bi-serial multiplier algorithm", IEEE Trans. on Computers, Vol. C-33, No. 10, pp. 906-911, 1984.

[8] H.M. Shao, T.K. Truong, L.J. Deutsch, J.H. Yuen and I.S. Reed, "A VLSI design of a pipeline Reed-Solomon Decoder", IEEE Trans on Computers, Vol. C-34, No. 5, pp. 393-403, 1985.

[9] B.B. Zyablov and E.L. Blokh, "Linear concatenated codes", (in Russian), Moscow 1982.

[10] S.M. Reddy and J.P. Robinson, "Random error and burst correction by iterated codes", IEEE Trans. on Inform. Theory, IT-18, No. 1, pp. 182-185, 1972.

[11] E.J. Weldon, "Decoding binary block codes on Q-ary output channels", IEEE Trans. on Inform. Theory, IT-17, No. 6, pp. 713-718, 1971.

[12] J. Justesen, "A class of constructive asymptotically good algebraic codes", IEEE Trans. on Inform. Theory, IT-18, No. 5, pp. 652-656, 1972.

[13] J.H. van Lint, "Error correction for concatenated codes", Nat. Lab. Report 1987 (unpublished).

$$
m \xrightarrow{\ f\ }
\begin{bmatrix} x_1 \\ x_2 \\ \\ x_a \end{bmatrix}
\begin{array}{l} \longrightarrow \\ \longrightarrow \\ \\ \longrightarrow \end{array}
\begin{bmatrix}
u_{11} & u_{12} & u_{1b} \\
u_{21} & u_{22} & u_{2b} \\
& & \\
u_{a1} & u_{a2} & u_{ab}
\end{bmatrix}
$$

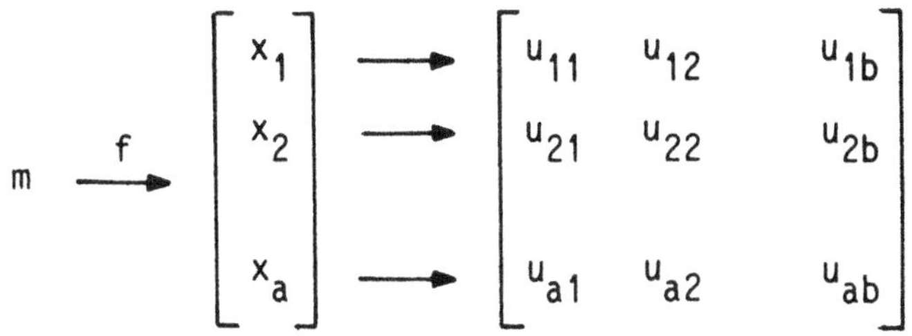

Fig. 1 The principle of concatenated encoding.

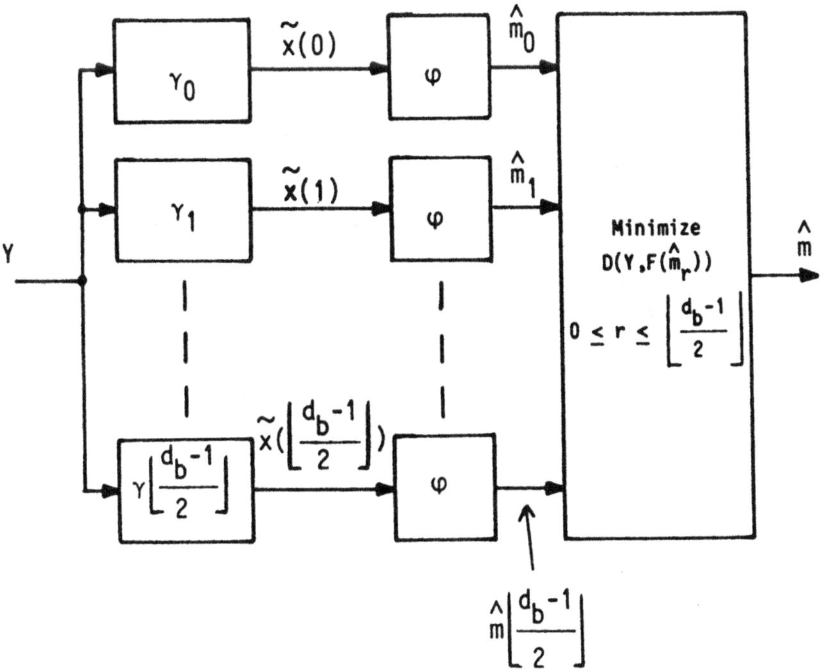

Fig. 2 The Blokh-Zyablov decoder.

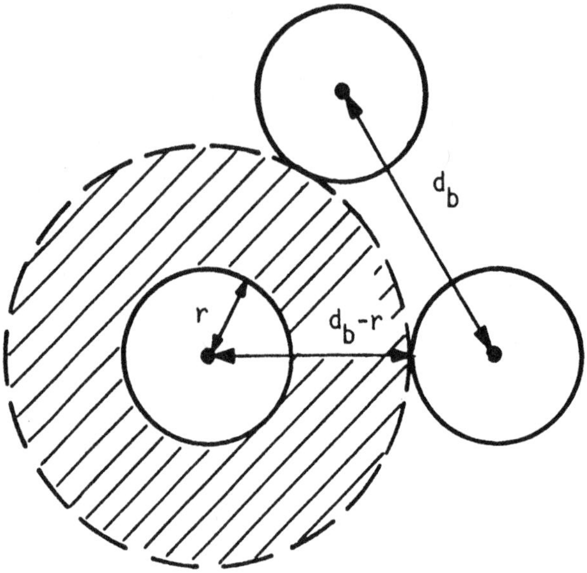

Fig. 3 Illustration to the inequality $\sum_{i=1}^{a} E(w_i,r) \geq 2t_r + P_r$. Decoding regions for the inner decoder γ_r are depicted. Supposing the lower left codeword is transmitted, $E(w_i,r)$ takes the value 0 within the inner circle; the value 1 in the dashed region, and 2 outside the dashed region.

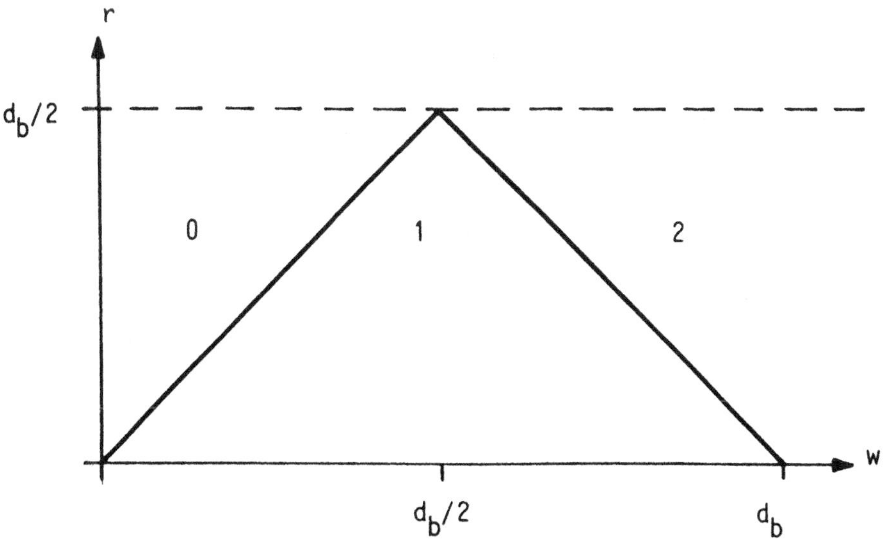

Fig. 4 Illustration to the equality $\int_0^{d_b/2} E(w,r)dr = \min\{w,d_b\}$.

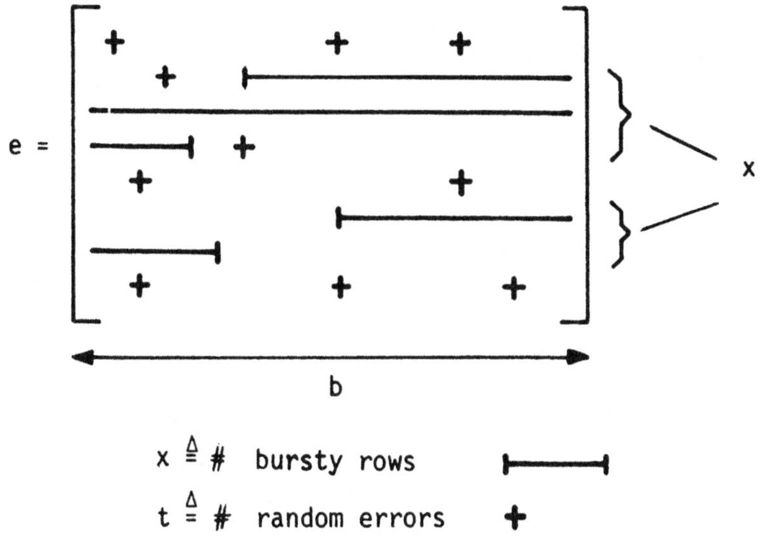

Fig. 5 Error matrix with bursts and random errors.

DESIGN OF A VITERBI DECODER WITH MICROPROCESSOR-BASED SERIAL IMPLEMENTATION

F.J. García-Ugalde, R.H. Morelos-Zaragoza A.
División de Estudios de Posgrado
Facultad de Ingeniería, UNAM
A.P. 70-256 Cd. Universitaria
04510 MEXICO, D.F.

ABSTRACT

The purpose of this paper is to present the design of a Viterbi decoder, for moderate data transmission rates (hundreds of bits/sec), using a serial implementation based on a 16/32-bit microprocessor.

This design is only one experimental phase of a final version which will be constructed to operate at a data transmission rate of 32 Kbits/sec, utilizing principally MECL and TTL integrated circuits.

1. INTRODUCTION

The Viterbi decoding algorithm has been widely used in communication systems. As a maximum-likelihood decoding scheme, it is the most practical forward-error-correcting (FEC) technique to achieve long coding gains in Gaussian channels, specifically in satellite channels [1].

In Mexico, with the place in orbit of the "Morelos" mexican domestic satellites, there is a plan to integrate data transmission, telephony and TV services with a KU-band earth stations network, throughout the country.

As a first step of this work, it started with the study of several channel coding schemes only for data transmission at a very low rate, which could be used in the above stations in its experimental phase. As a result of the present study it was stated that convolutional codes with Viterbi decoding give excellent performance, with relatively low complexity. Furthermore Viterbi decoders can easily be adapted to take advantage of soft demodulator decisions which will be a possibility in the final version of this error control system.

Only convolutional codes are considered here. Block codes which were common in early applications of digital space communication are so definitely inferior both in required complexity and in resulting performance that their further treatment is not worthwhile for the systems under consideration, other than as outer codes in a concatenated coding system.

2. GENERALITIES

To clarify some concepts used in this paper, we next present briefly the well known structure of convolutional codes and describe the Viterbi decoding algorithm, which is well known too [2] .

Figure 1 shows a coder for a binary convolutional code of rate m/n and constraint lenght k. The information bits are fed in m-bit shifts. The code constraint lenght is defined as the number of information bits that influence the n output channel bits.

The shift register is initially loaded with zeros, after which the information message is coded, and followed by a "tail" of k-m zeros to resynchronize the coder.

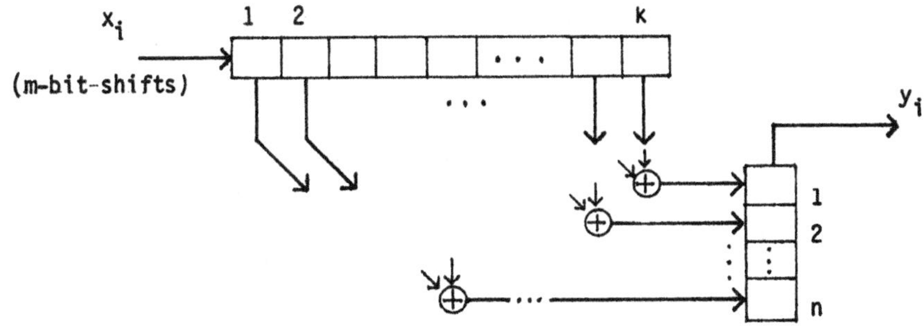

Fig. 1 A coder for a convolutional code of rate m/n and constraint lenght k.

To generate a n channel bits output, the coder considers m input information bits and the preceding k-m information bits.

Hence the memory of the coder is k-m. Since the coder is a lineal and time-invariant circuit, its state is given by the contents of the k-m memory elements, and together with the m input bits uniquely specify the output bits.

The structure of a binary convolutional code allow us to draw the output bit sequence using a trellis diagram shown in Fig. 2., for the 1/2-rate k=3 code. The trellis diagram is a transition diagram between the $2^{(k-m)}$ states of the coder. The branches indicate the state transition due to an input bit. Conventionally, a "zero" input results on an upper branch, whereas a "one" input results on a lower branch. The output is indicated above the corresponding branch.

The Viterbi decoding algorithm takes advantage of the repetitive structure of the trellis diagram.

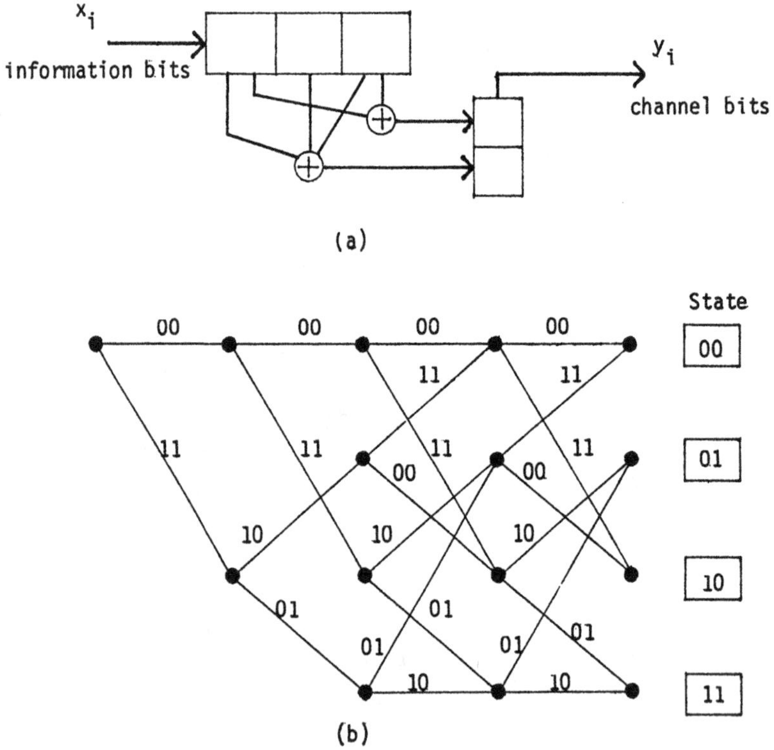

Fig. 2 (a) Coder for the convolutional code of rate 1/2 and constraint lenght k=3.
(b) Trellis diagram.

From the trellis diagram we observe that it has a repetitive structure after k-m consecutive bits.

The Viterbi decoder computes the distance to the received path of each of 2^m paths that arrive to a given state, and selects that path with maximum-likelihood, eliminating all other paths. This procedure is done for each state, in a given trellis depth. This is known as a decoding step. See Fig. 3 for the block diagram of such algorithm.

When eliminating the less likely paths that arrive to a state, the Viterbi decoder does not reject branches that might be part of the transmitted code path.

To compare paths entering a given state, we use the path-likelihood function given in [3] .

$$P(\bar{z}|\bar{Y}^{(m)}) \tag{1}$$

where $\bar{z}$ is the total received path and $\bar{Y}^{(m)}$ is one of the possible transmitted paths.

Let us consider the binary symmetric channel (BSC). The difference between the N-bit $\bar{z}$ and $\bar{y}^{(m)}$ paths is d_m bits (where d_m is the Hamming distance). The probability of $\bar{Y}^{(m)}$ being transformed into $\bar{z}$ is given by:

$$P(\bar{z}|\bar{Y}^{(m)}) = p^{d_m}(1-p)^{N-d_m} \tag{2}$$

where p is the BSC transition probability.

Using a log-likelihood function we have:

$$\log \left[P(\bar{z}|\bar{Y}^{(m)})\right] = - d_m \log \left[(1-p)/p\right] + N \log (1-p) \tag{3}$$

it is clear from (3) that in order to minimize the error probability in a BSC, we must choose the code path with minimum Hamming distance from the received path.

As for the memory required to store the history of the selected paths in each state (survivors), it has been shown [3] that the $2^{(k-m)}$ survivor paths tend to be equal in the preceding levels of the trellis, approximately 5k levels away from the present level (for 1/2-rate codes).

Therefore, the amount of memory required is about $5k \cdot 2^{(k-m)}$ bits, in addition to the one needed to store the distance of the survivor paths, which is of $5 \cdot 2^{(k-m)}$ bits for 1/2-rate codes.

This path memory truncation, results on avoiding the tail of the paths and permits a continuous operation, with negligible degradation in performance.

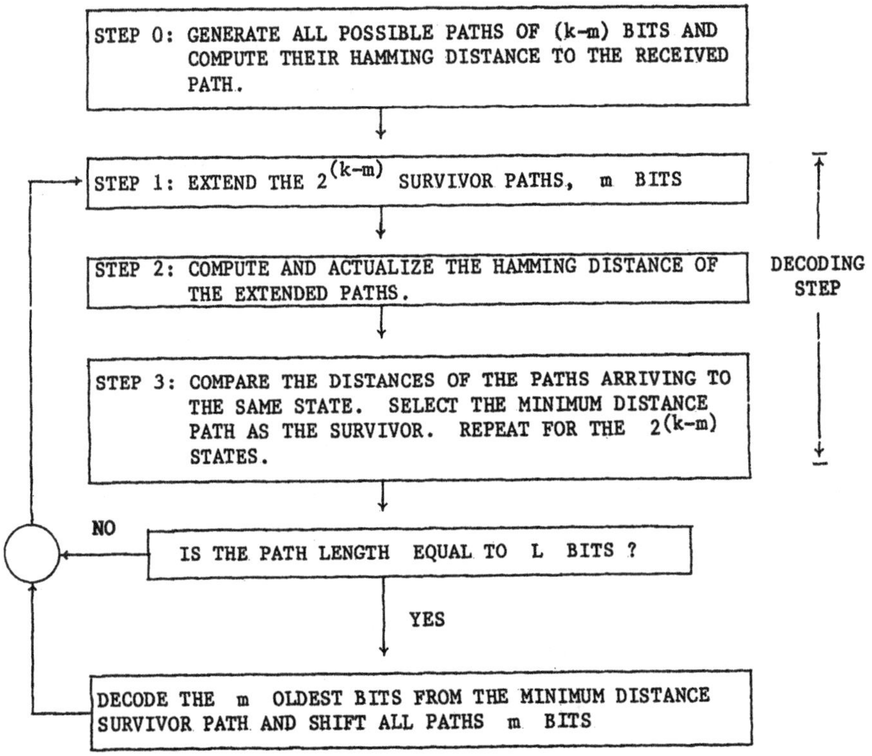

Fig. 3 Block diagram of the Viterbi decoding algorithm. The decoding lenght is L bits. k bits constraint lenght and m/n-rate code.

3. CONVOLUTIONAL CODE SELECTED

A 1×10^{-7} bit error rate (BER) corresponding to a theoretical 7.3 dB energy per bit-to-noise ratio (E_b/No) (not includes 0.5 dB margin), is one of the requirements for KU-band data link. In order to accomplish this, a 1/2-rate k=7 convolutional code was selected using a hard-decision Viterbi decoder. The code generators are those obtained by

Odenwalder in [4] , whose generator coefficients are 171, $133|_8$ = = 1111001, 1011011. These code generators were chosen to minimize the bit probability of error at large E_b/No ratios, which is the range of E_b/No's used here. These yields a code with a minimum free distance of 10.

In this version with hard-decisions, there is no need neither to compute more complicated likelihood functions, nor to use soft-decisions (usually of 3 bits) from the demodulator.

The choise of a rate-1/2 code allow us to use the availability of moderate bandwidth expansion, due to power limitation in the satellite. This rate also reduces significantly the decoder complexity, and we can exploit the fact that from this scheme, higher codes rates could be obtained with punctured codes [5] .

As for decoder's architecture two approximations were analyzed: parallel implementation and serial implementation.

With parallel implementation, $2^{(k-m)}$ processing units are needed for each state (each unit performing the operations of likelihood metric computation, comparing, adding, selecting and updating the survivor path). Utilizing principally MECL and TTL integrated circuits, this approach achieve very high data transmission rates (up to 10 Mbits/sec), at the cost of large number of logic and memory elements. This scheme is suitable for VLSI designs, obtaining a whole decoder in a single integrated circuit, whereas traditional electronic designs would require excessive space and would be impractical.

On the other hand, the serial approach permits significant saving in the number of devices required, but at the cost of reducing transmission rate (several hundreds of Kbits/sec). In this kind of scheme a single processing unit is required for all the $2^{(k-m)}$ survivor paths. In this version we consider the serial implementation as the most suitable for satellite data communications, because the present needs in data transmission rates are not to large -32 Kbits/sec using a SCPC/ QPSK/FDMA system, only for data transmission, nor telephony and TV services neither-, and also because it is a scheme easy to build with reduce number of components.

This type of architecture also has the advantage of being able to increase data transmission rate, by means of hybrid configuration - serial and parallel-, in which several processing units share the decoding task. The progressive decrement in both price and size of electronic devices is a valid justification of our selection.

4. PRESENT VERSION OF THE DESING

This preliminary version has been designed around a microprocessor for two reasons principally:

* To measure the performance of a relatively recent 16/32-bit microprocessor in this kind of applications.

* To test the correct operation of the decoding system accordingly to our theoretical calculations. Awaiting for the final version.

The key element in a serial implementation is the processing unit (PU), because it must perform the Viterbi decoding algorithm efficiently. High operation speed; efficiency in data addressing and transfer; and good instruction set are desirable features of the PU.

In the last few years more powerful microprocessors have appered, among which we find Motorola's MC68000 series that present all characteristics mentioned above. We therefore select a microprocessor-based serial implementation using the MC68000 family.

Being the microprocessor the heart of the design, the architecture must also consider three basic elements: a read-only-memory (ROM), to store the set of instructions to implement the Viterbi decoding algorithm; a random-access-memory (RAM), for the path distance storage and the path memory; and the input (from the demodulator) and output (to the user) interface circuits. The block diagram of the proposed design is shown in Fig. 4.

The system was constructed as a single board utilizing an Educational Computer Board (ECB) from Motorola systems. The principal features of this board are:

* A 16/32-bit microprocessor in its version of 10 MHz and N-channel technology.
* 32K bytes of dynamic RAM
* 8K bytes of system monitor ROM
* Two serial I/O ports with RS-232C interface. Baud rate, strap selectable: 110, 150, 300, 600, 1200, 2400, 4800, 9600.

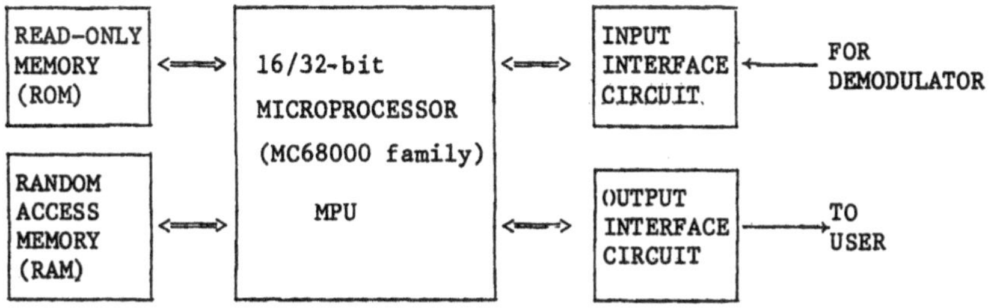

Fig. 4 Block diagram of the microprocessor-based serial implementation of Viterbi Decoder.

Therefore our choice is software implementation, using an assembler language program executed by the microprocessor. In a straight-forward implementation two sets of registers are used, containing the survivor paths and their distances, for two consecutive levels in the trellis diagram. After each decoding step, the microprocessor updates and transfers the contents of the registers sets.

The disadvantage of this procedure, as was pointed out in [6], is the reading and rewriting of long registers (32 bits in our case), which requires two buffers. Nevertheless, the speed and power of the new generation of microporcessors allow us to design without considering any restriction in memory size. Therefore, there is no need to remove double buffers for transfers between registers.

A very important feature of any decoder is the synchronization of symbols. In our case, branch synchronization is achieved by detecting when received data contain excessive number of errors. Whenever this happens, it can easily be detected by the microprocessor who shifts one bit the received branches.

To update and store the path distances we take advantage of the trellis structure, to determine state transitions [7] .

The basic trellis structure is used when comparing and updating both the survivor paths and their distances, as shown in Fig. 5.

In each decoding step, a new bit is added at the beginning of each path, and the oldest bit of the minimum distance path is decoded. In our

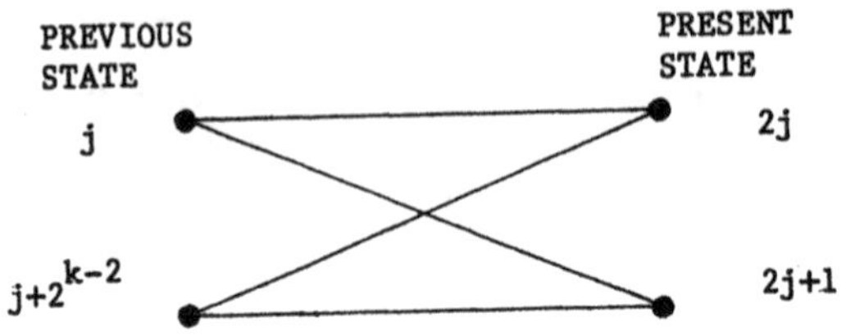

Fig. 5 Basic structure of the trellis diagram.

design, a path history of 31 bits was taken, and the 32nd bit is the decoded bit. At the end of the decoding step, the content of the registers is shifted right one bit.

With this implementation, nevertheless we could obtain a maximum data transmission rate of 495 bits/sec, which is about ten times faster than a previous work in which the Viterbi algorithm has been implemented on a Zilog Z-80 microcomputer for the (2,1)8 convolutional code [8].

Until now our design has been tested in our laboratories with a simulated channel and data transmission; the results are very encouraging, but much research still has to be done for this application, in particular with respect to the real data transmission rate (32 Kbits/sec), which will be obtained only with a logic design utilizing MECL and TTL integrated circuits or a VLSI realization.

5. CONCLUSION

In summary, a 16/32-bit microprocessor-based serial implementation of a Viterbi decoder has been designed. The scheme has the advantage of requiring a reduced number of components, and permits an increased data transmission rate by a serial-parallel approach. This work was done in the context of the design of KU-band earth stations for data transmission, telephony and TV services, using the mexican domestic satellites systems.

Nevertheless this design constitutes only one experimental phase of the final version, and although we could obtain a data transmission rate of 495 bits/sec which is about ten times faster than a previous work, it is not satisfactory for our present requirements which are of 32 Kbits/

sec for the data transmission link. To obtain this final rate a definitive version will be constructed utilizing MECL and TTL integrated circuits or a VLSI realization.

For the reason that has been explained in paragraph 1, only convolutional codes have been considered here, a 1/2-rate k=7 convolutional code has been chosen to accomplish one of the requirements for KU-band data link. To choose the code generators we have been highly influenced by the CCSDS recommendation for telemetry channel coding.

ACKNOWLEDGMENT

The authors are very grateful to the referees for valuable comments.

References

[1] J.A. Heller and I.M. Jacobs, "Viterbi Decoding for Satellite and Space Communications", IEEE Trans. on Commun. Technol., Vol. COM-19, No. 5, Oct. 1971, pp. 835-848.

[2] A.J. Viterbi, "Error Bounds for Convolutional Codes and an Asymptotically Optimum Decoding Algorithm", IEEE Trans. Inf. Theory, Vol. IT-13, Apr. 1967, pp. 260-269.

[3] A.J. Viterbi, "Convolutional Codes and Their Performance in Communication Systems", IEEE Trans. on Commun. Technol., Vol. COM-19, No. 5, Oct. 1971, pp. 751-772.

[4] J.P. Odenwalder, "Optimal Decoding of Convolutional Codes", Ph.D. thesis, University of California, Los Angeles, 1970.

[5] J.B. Cain, G.C. Clark Jr., and J.M. Geist, "Punctured Convolutional Codes of rate (n-1)/n and Simplified Maximum Likelihood Decoding", IEEE Trans. Inf. Theory, Vol. IT-25, No. 1, Jan. 1979, pp. 97-100.

[6] C.M. Rader, "Memory Management in Viterbi Decoder", IEEE Trans. on Commun., Vol. COM-29, No. 9, Sep. 1981. pp. 1399-1401.

[7] G.D. Forney, "The Viterbi Algorithm", Proc. IEEE, Vol. 61, Mar. 1973, pp. 268-277.

[8] H.H. Ma, "The Multiple Stack Algorithm Implemented on a Zilog Z-80 Microcomputer", IEEE Trans. on Commun., Vol. COM-28, No. 11, Nov. 1980, pp. 1876-1882.

ON s-SUM-SETS (s odd) AND THREE-WEIGHT PROJECTIVE CODES

Mercè GRIERA

Departament d'Informàtica, Facultat de Ciències,
Universitat Autònoma de Barcelona, Bellaterra,
Barcelona, Espanya.
and
G.E.C.T., Université de Toulon et du Var
83130 La Garde, France.

We show that if Ω is the set of coordinate forms of a three-weight linear projective code $C(n,k)$, and if $X=F^*\Omega$ is a 5-sum-set then the three weights are in arithmetical progression, that is, $w_1= w_2 - A$ and $w_3= w_2 + A$ with A a function which depends of the number of words of weight three in $C^{\perp}(n,n-k)$. Furthermore we obtain some relations between s-sum-sets (s odd) and their parameters.

I.- INTRODUCTION

Let Ω be the set of coordinate forms of a linear projective code C(n,k) and $X=F^*\Omega$. In (4) we show that if X is an s-sum-set with s even then C(n,k) has not more than two non-zero distinct weights, that is, X is a partial-difference-set in the sense of Camion (2).

We also show that if s is odd then C(n,k) has not more than three non-zero distinct weights, but we do not show any relation with the triple-sum-sets in the sense of Courteau and Wolfmann (3).

Therefore, in this paper, we only consider s-sum-sets with s odd and three-weights projective codes.

First we characterize s-sum-sets X, coming from a set of coordinate forms of a three-weights projective code, in polynomial form, concretely by a Diophantine equation. We then show that if X is a 5-sum-set then X is an s-sum-set for all $s=2t+1\geq 3$ ($t \geq 1$) and the parameters are obtained as a function of parameters of the 3-sum-set.

For the reciprocal we need to add in the hypothesis that X be a 3-sum-set, therefore there are two open questions: does there exist a 3-sum-set which is not s-sum-set for all $s\neq 3$? and, does there exist an s-sum-set for any $s \geq 7$ which is not 3-sum-set?.

We also show that if X is a 5-sum-set then the three non-zero distinct weights have the following structure:

$$w_1 = (\text{card } X / q) - A \qquad w_2 = \text{card } X / q \qquad w_3 = (\text{card } X / q) + A$$

with A a function of n,k,q and γ_3, where γ_3 is a parameter related to the number of words of weight three in $C^{\perp}(n,n-k)$.

This paper owes a lot to J. Wolfmann's ideas.

II.- NOTATION AND PREVIOUS DEFINITIONS

Let F=GF(q) be the q element Galois field, q a power of a prime p, and let $F^* = F \setminus \{ 0 \}$.

II.1.- DEFINITION

A subset $X \subset F^k$ is said to be an s-sum-set with parameters λ_s and μ_s if :

a) $F^*X = X$

b) $$\text{card}\{(x_1,x_2,\ldots,x_s) \in X^s \;/\; x = \sum_{i=1}^{s} x_i\} = \begin{cases} \lambda_s \text{ if } x \in X \setminus \{0\} \\ \mu_s \text{ if } x \in X^C \setminus \{0\} \end{cases}$$

where $X^C = F^k \setminus X$ is the complement of X.

II.2.- DEFINITION

If X is a s-sum-set we define the parameter γ_s as follows:

$$\gamma_s = \text{card}\{(x_1,x_2,\ldots,x_s) \in X^s \;/\; \sum_{i=1}^{s} x_i = 0\}$$

II.3.- DEFINITION

Let $X \subset F^k$ be a subset such that $F^*X = X$ and rank $X = k$. X determines the following partition of F^k:

$x \in C_0$ iff $x=0$

$x \in C_i$ iff $x \in X+X+\ldots\ldots+X$ (i times) and $x \notin C_j \quad \forall\, j<i$

Thus $F^k = C_0 \cup C_1 \cup \ldots\ldots \cup C_r$ with $C_i \cap C_j = \varnothing \quad \forall\, i \neq j$.

Furthermore the above partition induces a translation invariant partition of $F^k x F^k$:

$F^k \times F^k = R_0 \cup R_1 \cup \ldots\ldots \cup R_r$ where $(x,y) \in R_i$ iff $y-x \in C_i$.

II.4.- DEFINITION

If we consider the characteristic functions of each relations R_i, then the endomorphisms, called adjacencies, in $C[F^k]$ are obtained as follows:

$$D_i(x,y) = \begin{cases} 1 \text{ if } y-x \in C_i \\ 0 \text{ otherwise} \end{cases}$$

Clearly $D_0 = Id$ and $D_0 + D_1 + \ldots\ldots + D_r = J$ where $J(x,y) = 1 \quad \forall\, (x,y) \in F^k x F^k$.

D_i can be expressed as a $q^k x q^k$ matrix indexed by the elements of F^k, with $D_i(x,y)$ denoting the element corresponding to the row x and column y.

III.-CHARACTERIZATION OF s-SUM-SETS

III.1.- LEMMA

Let E be a vectorial space of dimension n over K (a commutative field) and set $\varphi \in End(E)$ such that $Ker\,\varphi \cap Imag\,\varphi = 0$ and $dim(Ker\,\varphi) = 1$.

Set $H_\varphi = \{ \psi \in End(E) \ / \ \psi\varphi = \varphi\psi = 0\}$.

Then H_φ is a subspace of dimension 1.

Proof:

We have $E = Ker\,\varphi \oplus Imag\,\varphi = \langle u \rangle \oplus \langle v_1, v_2, \ldots, v_{n-1} \rangle$ where $\{v_1, v_2, \ldots, v_{n-1}\}$ is a basis of Imag φ.

If $\psi \in H_\varphi$, then $\psi / _{Imag\,\varphi} = 0$ and $\psi(u) = \lambda_1 v_1 + \lambda_2 v_2 + \ldots + \lambda_{n-1} v_{n-1} + \mu u$

Hence $0 = \varphi(\psi(u)) = \varphi(\lambda_1 v_1 + \lambda_2 v_2 + \ldots + \lambda_{n-1} v_{n-1})$ since $\varphi(u) = 0$

Thus, $\lambda_1 v_1 + \lambda_2 v_2 + \ldots + \lambda_{n-1} v_{n-1} \in Ker\,\varphi \cap Imag\,\varphi$ and $\lambda_i = 0 \quad \forall i$.

Therefore $\psi(u) = \mu u$ and $\psi(Imag\,\varphi) \equiv 0$ and the lemma is proved.

III.2.- LEMMA

Let $\xi \in End(E)$ such that $Ker\,\xi \cap Imag\,\xi = 0$ and $dim(Imag\,\xi) = 1$.

If $\psi \in H_\xi$ and $dim(Ker\,\psi) = 1$

then $Imag\,\psi = Ker\,\xi$ and $Imag\,\xi = Ker\,\psi$.

Proof:

We have $E = Ker\,\xi \oplus Imag\,\xi = \langle v_1, v_2, \ldots, v_{n-1} \rangle \oplus \langle u \rangle$ where $\{v_1, v_2, \ldots, v_{n-1}\}$ is a basis of Ker ξ.

If $\psi \in H_\xi$ $\psi(u) = 0$, that is, $u \in Ker\,\psi$ and $Ker\,\psi \supset Imag\,\xi$, but they have the same dimension, then we obtain $Ker\,\psi = Imag\,\xi$. (iii)

If $\psi \in H_\xi$ $0 = \psi(\xi(v_i)) = \xi(\psi(v_i))$, that is, $\psi(v_i) \in Ker\,\xi$.

On the other hand $\psi(u) \in Ker\,\xi$ since $\psi(u) = 0$. Then $Ker\,\xi \supset Imag\,\psi$ and by (iii) $Imag\,\psi = Ker\,\xi$.

III.3.- THEOREM

Let $X \subset F^k$ be a set such that $F^*X = X$ and rank $X = k$. Set $X' = X \setminus \{0\}$.
If $dim(Ker(D_1 - card\,X'Id)) = 1$ then:

A.- If $0 \notin X$, the following two statements are equivalent:

i) X is an s-sum-set

ii) $\exists\ \delta_s, \varepsilon_s \in Z$ such that $(D_1 - \text{card } X Id)(D_1{}^s + \delta_s D_1 + \varepsilon_s) = 0$

B.- If $0 \in X$, the following two statements are equivalent:

i) X is an s-sum-set

ii) $\exists\ \delta_s, \varepsilon_s \in Z$ such that $((D_1+Id) - \text{card } X Id)((D_1+Id)^s + \delta_s(D_1+Id) + \varepsilon_s) = 0$

<u>Proof:</u>

ii) ==> i)

A.-

It is easy to prove that the endomorphism J, such that $J(x,y)=1\ \forall\ x,y$ satisfies the hypothesis of lemma III.2; in fact:

1.- $u \in \text{Ker } J \cap \text{Imag } J \implies \exists\ a\ /\ J(a)=u$ and $J(J(a))=0 \implies q^k J(a)=0 \implies J(a)=u=0$.

2.- $\dim(\text{Imag } J) = \text{rank } J = 1$.

On the other hand, the endomorphism $(D_1 - \text{card } X\ Id) \in H_J$ because:

$$(D_1 - \text{card } X\ Id)\ J = J\ (D_1 - \text{card } X\ Id) = D_1 J - \text{card } X\ J = 0$$

Here $X' = X$, so by lemma III.2 we have:

$$\text{Ker } (D_1 - \text{card } X\ Id) = \text{Imag } J$$
$$\text{Imag } (D_1 - \text{card } X\ Id) = \text{Ker } J$$

Therefore $\text{Ker } (D_1 - \text{card } X\ Id) \cap \text{Imag } (D_1 - \text{card } X\ Id) = \text{Ker } J \cap \text{Imag } J = 0$.

Thus $(D_1 - \text{card } X\ Id)$ verifies the hypothesis of lemma III.1 and we can conclude that $\dim(H_{(D1 - \text{card } X\ Id)}) = 1$.

Because $J, (D_1{}^s + \delta_s D_1 + \varepsilon_s) \in H_{(D1 - \text{card } X\ Id)}$ by hypothesis, there exist ρ_s such that $(D_1{}^s + \delta_s D_1 + \varepsilon_s) = \rho_s J$.

Using theorem II.5 in (4) and taking $\mu_s = \rho_s$ $\lambda_s = \rho_s - \delta_s$ $\alpha_s = \rho_s - \varepsilon_s$ we conclude that X is an s-sum-set.

B.-

This case needs the same method but with J and $((D_1 + Id) - \text{card } X\ Id) = (D_1 - \text{card} X'\ Id)$. We show that $((D_1+Id)^s + \delta_s (D_1+Id) + \varepsilon_s) = \rho_s J$.

Using theorem II.5 from (4) and taking $\mu_s = \rho_s$ $\lambda_s = \rho_s - \delta_s$ $\alpha_s = \rho_s - \varepsilon_s - \delta_s$ we conclude that X is an s-sum-set.

i) ==> ii)

In both cases (A.- and B.-) is the corollary II.6 of (4).

<u>IV.- s-SUM-SETS AND PROJECTIVE CODES</u>

<u>IV.1.- PROPOSITION</u>

Let Ω be the set of coordinate forms of $C=C(n,k) \subset F^n$ a projective linear code generated by an (k x n) matrix G and set $X = F^*\Omega \subset F^k$, then:

$$\dim(\operatorname{Ker}(D_1 - \operatorname{card} X\ \mathrm{Id})) = 1$$

Proof:

Let $\hat{F^k}$ be the group of characters of F^k and for each $x \in F^k$ let f_x denote the corresponding element of the dual basis. If $\chi \in \hat{F^k}$ then

D_1 diagonalizes in the characters basis and $\sum_{u \in F^k} \chi(u) f_u$ is an eigenvector with eigenvalue $\sum_{x \in X} \chi(x)$.

Therefore $\sum_{x \in X} \chi(x) = \operatorname{card} X$ iff $\chi(x) = 1 \quad \forall x \in X$.

Since F^k is additively generated by X, we conclude that $\chi(u)=1 \quad \forall u \in F^k$.

Hence, $\sum_{u \in F^k} f_u$ generates the eigensubspace associated to the eigenvalue card X of D_1.

IV.2.- THEOREM

Let Ω be a set of coordinate forms for a three-weight linear projective code $C=C(n,k)$; w_1, w_2, w_3 being the three non-zero weights.

Set $X=F^*\Omega$, $X'=X \cup \{0\}$ and for $i=1,2,3$ $\alpha_i= n(q-1)-qw_i$ $\alpha'_i= 1+n(q-1)-qw_i$

Then:

A.- X is an s-sum-set iff $P_{s-2}(\alpha_1, \alpha_2, \alpha_3) = 0$

B.- X' is an s-sum-set iff $P_{s-2}(\alpha'_1, \alpha'_2, \alpha'_3) = 0$

where $P_h(x,y,z) = 0$ is the Diophantine equation $\sum x^i y^j z^k=0$ the sum being over triples (i,j,k) such that $0 \le i, j, k \le h$ and $i + j + k = h$.

Proof:

A.-

Using theorem III.3 and the expression of the annihilator polynomial of D_1 (see theorem III.5 in (4)) we can characterize s-sum-sets X that proceed from three weights projective linear codes in the following form:

X is an s-sum-set iff $\exists\ \delta_s, \varepsilon_s$ such that $(Z-\alpha_1)(Z-\alpha_2)(Z-\alpha_3)$ divides $Z^s + \delta_s Z + \varepsilon_s$

Set $A = -(\alpha_1 + \alpha_2 + \alpha_3)$ $B = \alpha_1\alpha_2 + \alpha_1\alpha_3 + \alpha_2\alpha_3$ $C = -\alpha_1\alpha_2\alpha_3$ then

X is an s-sum-set iff $\exists\ \delta_s, \varepsilon_s$ such that $Z^3 + A Z^2 + B Z + C$ divides $Z^s + \delta_s Z + \varepsilon_s$

The case s=3 is easily proved since:

$$Z^3 + A\ Z^2 + B\ Z + C = Z^3 + \delta_3\ Z + \varepsilon_3 \quad \text{iff} \quad A=0 = P_1(\alpha_1, \alpha_2, \alpha_3)\ , \quad B = \delta_3 \quad \text{and} \quad C = \varepsilon_3$$

For s>3

$$Z^s + \delta_s Z + \varepsilon_s = (Z^3 + A\ Z^2 + B\ Z + C)\,(Z^{s-3} + a_1 Z^{s-4} + \dots\dots\dots + a_{s-3})$$

if we equalize the coefficients we obtain the following s equations:

$$\begin{array}{l} A + a_1 = 0 \\ B + Aa_1 + a_2 = 0 \\ C + Ba_1 + Aa_2 + a_3 = 0 \\ \dots\dots\dots\dots\dots\dots\dots\dots \\ \dots\dots\dots\dots\dots\dots\dots\dots\dots\dots \\ Ca_{s-6} + Ba_{s-5} + Aa_{s-4} + a_{s-3} = 0 \\ Ca_{s-5} + Ba_{s-4} + Aa_{s-3} = 0 \\ Ca_{s-4} + Ba_{s-3} = \delta_s \\ Ca_{s-3} = \varepsilon_s \end{array}$$

We can write the first s-3 equations in a recurrent form:

$$(\theta^3 + A\theta^2 + B\theta + C)\,(a_h) = 0 \qquad h = 0, \dots\dots, s-6$$

with initial values:

$$\begin{array}{l} a_0 = 1 \\ a_1 = \alpha_1 + \alpha_2 + \alpha_3 \\ a_2 = \alpha_1^2 + \alpha_2^2 + \alpha_3^2 + \alpha_1\alpha_2 + \alpha_1\alpha_3 + \alpha_2\alpha_3 \end{array}$$

Then X is an s-sum-set iff $Ca_{s-5} + Ba_{s-4} + Aa_{s-3} = 0$, that is $a_{s-2} = 0$ and we have the parameters in the two last equations.

It is well known that $a_h = c_1\alpha_1^h + c_2\,\alpha_2^h + c_3\,\alpha_3^h$ where the coeficients c_1, c_2, c_3 are determined by the initial values. This gives:

$$a_h = \frac{\alpha_1^{h+2}}{(\alpha_2 - \alpha_1)(\alpha_3 - \alpha_1)} - \frac{\alpha_2^{h+2}}{(\alpha_3 - \alpha_2)(\alpha_2 - \alpha_1)} + \frac{\alpha_3^{h+2}}{(\alpha_3 - \alpha_1)(\alpha_3 - \alpha_2)}$$

Reducing to the same denominator and dividing, we obtain $a_h = P_h(\alpha_1, \alpha_2, \alpha_3)$ and the theorem follows since the B.- case is treated in the same fashion.

IV.3.- COROLLARY

With the same hypothesis as in the theorem, if X (or X') are 5-sum-sets then X (or X') are s-sum-sets for all $s = 2j+1$, $j \geq 1$.

The parameters are related by:

$$\begin{cases} \mu_s - \lambda_s = (-1)^{j-1}\,(\mu_3 - \lambda_3)^j & s \geq 3 \\ m^s = \lambda_s m + (q^k - m)\,\mu_s & \end{cases}$$

where $m = n(q-1)$ in the X case and $m = 1 + n(q-1)$ in the X' case.

Proof:

If X is a 5-sum-set then $P_3(\alpha_1, \alpha_2, \alpha_3) = 0$. By (6) we know that $\alpha_i = 0$ for some i. If we suppose $\alpha_1 = 0$ then $\alpha_2 = -\alpha_3$ and $A = C = 0$.

Now, the above recurrent equation is $(\theta^3 + B\theta)(a_h) = 0$ with initial values $a_0 = 1$, $a_1 = 0$ and $a_2 = -B$. Solutions are $a_{2j+1} = 0$ and $a_{2j} = (-1)^j B^j$ for all $j > 0$, that is, X is an s-sum-set for all s odd.

If X is a 3-sum-set then $A = 0$, $B = \delta_3$ and $C = \varepsilon_3$. Furthermore if X is a 5-sum-set then $\varepsilon_3 = \varepsilon_5 = \varepsilon_7 = \ldots\ldots\ldots = \varepsilon_{2j+1} = 0$ and $Z^3 + \delta_3 Z$ divides $Z^s + \delta_s Z$, that is :

$$\delta_{2j+1} = (-1)^{j-1} B^j = (-1)^{j-1} \delta_3^j \quad j \geq 1.$$

Since $\delta_{2j+1} = \mu_{2j+1} - \lambda_{2j+1}$ we have:

$$\mu_{2j+1} - \lambda_{2j+1} = (-1)^{j-1} (\mu_3 - \lambda_3)^j \quad \forall j \geq 1.$$

For the second equation:

If we take the definiton of s-sum-set we can write:

$$\text{card } X^s = \gamma_s + \lambda_s \text{ card } (X \setminus \{0\}) + \mu_s \text{ card } (X^C \setminus \{0\}) .$$ Now:

In the X case $[n(q-1)]^s = \gamma_s + n(q-1)\lambda_s + [q^k - n(q-1) - 1]\mu_s$. Using $\varepsilon_s = 0$, that is, $\mu_s = \gamma_s$ we conclude that $[n(q-1)]^s = n(q-1)\lambda_s + [q^k - n(q-1)]\mu_s$.

In the X' case $[n(q-1)+1]^s = \gamma_s + n(q-1)\lambda_s + [q^k - n(q-1) - 1]\mu_s$. Using $\varepsilon_s = 0$, that is, $\lambda_s = \gamma_s$ we conclude that $[n(q-1)+1]^s = [n(q-1)+1]\lambda_s + [q^k - n(q-1) - 1]\mu_s$.

IV.4.- REMARK

Let Ω be a set of coordinate forms for a three-weights linear projective code $C = C(n,k)$; w_1, w_2, w_3 being the three non-zero weights. Set $X = F^*\Omega$, $X' = X \cup \{0\}$.

1.- X (or X') are 5-sum-sets if and only if they are 3-sum-sets and there is a weight, for instance w_1, such that $w_1 = \text{card } X / q$ (or $w_1 = \text{card } X'/q$).

2.- X (or X') are 5-sum-sets if and only if they are 3-sum-sets such that the zero doesn't have a particular treatment, that is if :

$$\text{card}\{(x_1,x_2,x_3) \in X^3 \ / \ x = \sum_{i=1}^{3} x_i\} = \begin{cases} \lambda_3 & \text{if } x \in X \\ \mu_3 & \text{if } x \in X^C \end{cases}$$

IV.5.- COROLLARY

With the same hypothesis as in the theorem, if X (or X') is a 3-sum-set and an s-sum-set for some $s \geq 7$ odd, then X (or X') are a 5-sum-set.

Proof:

It is easily shown that for each s odd we can write:

$$P_s(\alpha_1, \alpha_2, \alpha_3) = A_s(\alpha_1 + \alpha_2 + \alpha_3) + B_s(\alpha_1 \alpha_2 \alpha_3)$$

where A_s and B_s are sums of products of even powers of α_i. Now, the corollary is proved.

IV.6.- COROLLARY

With the same hypothesis as in the theorem, if X (or X') is a 5-sum-set then the three weights have the following structure:

A.- In the X case: If $m = n(q-1)$

$$w_1 = m/q - 1/q\left(\sqrt{(m^3-q^k\gamma_3)/m}\right) \qquad w_2 = m/q \qquad w_3 = m/q + 1/q\left(\sqrt{(m^3-q^k\gamma_3)/m}\right)$$

B.- In the X' case: If $m = 1 + n(q-1)$

$$w1 = m/q-1/q\left(\sqrt{(q^k\gamma_3 - m^3)/(q^k- m)}\right) \qquad w_2 = m/q \qquad w_3 = m/q + 1/q\left(\sqrt{(q^k\gamma_3 -m^3)/(q^k-m)}\right)$$

Proof:

By remark IV.4.-, X is a 3-sum-set. Thus $(z - \alpha_1)(z - \alpha_2)(z - \alpha_3)$ divides $Z^3+\delta_3 Z+\varepsilon_3 = Z^3+ (\lambda_3 - \mu_3) Z = Z(Z^2+ (\lambda_3 - \mu_3))$ and $\alpha_1, \alpha_2, \alpha_3$ are the zeroes of this polinomial. This gives

$$\alpha_2 = 0, \qquad \alpha_1 = \sqrt{\lambda_3 - \mu_3} \qquad \text{and} \qquad \alpha_3 = -\sqrt{\lambda_3 - \mu_3}$$

On the other hand $\alpha_i = n(q-1)-qw_i$ where w_1, w_2, w_3 are the non zero weigths of the code generated by X. Hence

$$w_1 = 1/q\left(n(q-1) - \sqrt{\lambda_3 - \mu_3}\right), \quad w_2 = n(q-1)/q, \quad w_3 = 1/q\left(n(q-1) + \sqrt{\lambda_3 - \mu_3}\right)$$

The X' case is treated in the same way.

Now by using corollary IV.3.- we may determine $\lambda_3 - \mu_3$ as follows

$$m^3 = \lambda_3 m + [q^k-m]\mu_3 = m(\lambda_3 - \mu_3) + q^k \mu_3 .$$

In the X case, $\mu_3 = \gamma_3$ and we have $\lambda_3 - \mu_3 = (m^3-q^k\gamma_3)/m$

In the X' case, $\lambda_3 = \gamma_3$ and we obtain $\lambda_3 - \mu_3 = (q^k\gamma_3 - m^3)/(q^k- m)$.

IV.7.- PROPOSITION

Let Ω be a set of coordinate forms for a three-weights linear projective code $C=C(n,k)$. Set $X=F^*\Omega$, $X'=X \cup \{0\}$ and B_3 the number of words of weight three in $C^{\perp}(n,n-k)$.

A.- If $\gamma_3 = \text{card}\{(x_1,x_2,x_3) \in X^3 / \sum_{i=1}^{3} x_i = 0\}$ then $\gamma_3 = 6B_3 + (q-1)(q-2)n$

B.- If $\gamma_3 = \text{card}\{(x_1,x_2,x_3) \in X'^3 / \sum_{i=1}^{3} x_i = 0\}$ then $\gamma_3 = 6B_3 + n(q^2-1) + 1$

Proof:

If $u = (u_1,u_2,\ldots\ldots,u_n) \in C^{\perp}(n,n-k)$ and $w(u) = 3$ $\exists\, u_i, u_j, u_l \in F^*$ such that $u_i\phi_i + u_j\phi_j + u_l\phi_l = 0$ for $\phi_i, \phi_j, \phi_l \in \Omega$, that is , for any word of weight three in C (n,n-k) we have 3! sums resulting zero of elements of X.

Because the code is projective the elements of Ω are pairwise linearly independent . Hence, if we take $x\phi_i + y\phi_i + z\phi_l = 0$ for $x, y, z \in F^*$ then $\phi_l = \phi_i$, $(x+y+z)\phi_i = 0$ and $x+y+z = 0$.

In the A.- case, since $0 \notin X$ there is only one possibility, that is, $y \neq -x$ and $z = -(x+y)$. Thus for any $\phi_i \in \Omega$ there are (q-1) (q-2) zero-sums of three elements of X.

In the B.- case, $0 \in X'$ We need to add to the A.- method the following possibility:

$x\phi_i + y\phi_j + z\phi_l = 0$ with x or y or z = 0, we suppose x.

Because the code is projective, in this case $\phi_j = \phi_l$ and $(y+z)\phi_j = 0$, that is, $z = -y$

$y = 0$ only represents a sum of elements of X': 0 + 0 +0

$y \neq 0$ represents three sums : $0 + y\phi_j + (-y)\phi_j$

$y\phi_j + 0 + (-y)\phi_j$

$y\phi_j + (-y)\phi_j + 0$

that is, for any $\phi_j \in \Omega$ there are 3(q-1) sums being zero and the lemma holds.

V.- REFERENCES

(1) BLAKE "The Mathematical Theory of Coding" Academic Press. New York (1975)

(2) CAMION P. "Difference sets in Elementary Abelian Groups" Les presses de l'Université de Montréal, Montréal. (1979)

(3) COURTEAU B. and WOLFMANN J. "On triple-sum-sets and two or three weights codes" Discrete math. 50 (1984)

(4) GRIERA M.and RIFA J. and HUGUET Ll. "On s-sum-sets and projective codes" Lecture Notes in Computer Sciences, vol 229 pag 135-142 (1986)

(5) GRIERA M. "Esquemes d'associació: Aplicació a teoria de codis" These : Universitat Autònoma de Barcelona. (1984)

(6) WARD M. "The vanishing of the homogeneus product sum of the roots of a cubic" Duke Math., J.26 , pag. 553-562. (1959)

ON THE INTEGRATION
OF
NUMERIC AND ALGEBRAIC COMPUTATIONS

Gianfranco Mascari

Istituto per le Applicazioni del Calcolo
Consiglio Nazionale delle Ricerche
Viale del Policlinico 137
00161 Roma, Italy

Alfonso Miola

Dipartimento di Informatica e Sistemistica
Università di Roma "La Sapienza"
Via Eudossiana 18
00184 Roma, Italy

This paper studies the properties of computations on numeric and algebraic expressions in order to capture their relations and differences. An analysis of the two kinds of computations from the point of view of program semantics is made. The possibility of integrating these two kinds of computations in a unified computational framework is discussed. The use of p-adic arithmetic, as a possible useful interpretation of algebraic expression, is considered to allow exact mathematical computations in that framework.

---

1. Introduction

The interface between numeric and algebraic computations is one of the biggest problem that researchers in Symbolic and Algebraic Computation (SAC) have faced from the beginning of their activities in the early sixties. Many scientists from that time have in fact looked at SAC as a drastic choice to be made in scientific computation only when a closed form solution to a given problem is absolutely necessary or when all the known numerical methods fail to answer the given question.

The wide and meaningful results of the last fifteen years in SAC give us now a different shape of the relations between numeric and algebraic computations. These two kinds of computations are currently shown to be allied in scientific problem solving. Survey papers [B-H79,E-M85,NGE79] present many interesting considerations on the current numeric and algebraic interface, together with a very useful set of references to relevant applications.

At the same time the necessity of defining the properties of numeric computing by a precise mathematical formulation has been recognized and some interesting proposals have already been made [ALB80,K-M80,BRO81,C-V85,BCR86] and work is in progress within the ESPRIT projects ATES and DIAMOND for the development of advanced numerical computing environments. Such a "synergism of the concepts and methods of computer algebra computations, and those which invoke approximations to real numbers and functions on them" has been recognized also in the Symposium on Accurate Scientific Computations [M-T85].

In order to clarify the relationship of the two kinds of computations we may characterize them by the following schema. The algebraic computing offers the possibility of carrying out computations in different domains. That is obtained basically because it guarantees the computation in abstract structures and the transformations and the extensions of these structures by well defined algebraic mechanisms.

Such mechanisms have been recognized to represent very powerful tools from a computational point of view [YUN76][M-Y74] and at the same time have been considered as generalizations of classical approximation methods [YUN75,LIP76].

On the other side the numeric computing has always been seen as a class of methods and tools for the evaluation and the approximation of various classes of functions in a fixed domain, namely the real numbers domain, expressed by computer arithmetics. However it must be underlined that such arithmetics suffer from strong qualitative limitations from the point of view of formal correctness.

The actual need is to give a unified view of those two computing frameworks in order to achieve the design of an integrated computing environment. To this purpose, two points of view can be followed, namely a pragmatic one based on the current and future possibilities of software environments supporting scientific computations, and a more theoretical one, based on the objective of defining a unique semantic description of the two kinds of computations. These two points of view mainly correspond to the user's needs interested in flexible interface and to the software system designers' interests to build systems of a completely new structure based on a solid methodology.

This paper considers the latter of those two directions. More precisely we refer to methodological achievements both from the program semantics and the computational constructions points of view, and to a new powerful class of tools, such as the SCRATCHPAD II system [BDG84], allowing formulation on different computing domains. Our main purpose is to consider both the semantic framework in which to formulate and analyse the integration problem and the computational paradigms to correctly offer a possible realization of the proposed approach.

2. Algebraic and Numeric Computations: Semantical Aspects

In this section we aim to verify the feasibility of the integration problem from a theoretical point of view. To this purpose we take the context of program semantics where it will be possible to precisely formulate the equivalence of the two kinds of computations by means of the known equivalence of operational and algebraic semantics of program schemes.

We are going to show this correspondence by recalling some basic definitions and results from algebraic semantics, just for clarity and completeness, full expositions can be found in [ADJ77,C-N76] for the approach based on complete partial orders and in [A-N80] for the approach based on complete metric spaces, which we shall consider.

Let us recall that by a *recursive program scheme* Σ on an alphabet F, i.e a finite set of function symbols with arity, we mean a set of equations

$$\Sigma = \{\Phi_i(x_1, \ldots, x_{n_i}) = \tau_i | i = 1, \ldots, k\}$$

where $\Phi = \{\Phi_1, \ldots, \Phi_k\}$ is a set of unknown function symbols, $\Phi \cap F = \emptyset$, $\{x_1, \ldots, x_{n_i} \mid i = 1, \ldots, k\}$ is a set of variables, τ_i is a term in $T(F \cup \Phi, \{x_1, \ldots, x_{n_i}\})$, the free $(F \cup \Phi)$-algebra generated by $\{x_1, \ldots, x_{n_i}\}$.

The intuitive idea which underlies the notion of *result of a successful computation* is the following one "..., at the beginning of the computation we only know that the value of the computed function domain and in the course of the computation this range is reduced", [A-N80].

This kind of semantic interpretation allows also to treat in a satisfactory way the results of infinite computations.

2.1. Numeric computations and operational semantics

Operational semantics consists in specifying a complete metric F-algebra as interpretation of an alphabet, and defining the result of a successful computation from an interpreted term with a program, i.e. a program scheme and an interpretation. Numeric computations can be seen as those successful computations from a term t (the expression to evaluate), a program scheme Σ and an interpretation I consisting (in most cases) of the complete metric field $\mathcal{R}$ of real numbers. A *complete metric F-algebra* is a structure $I = \langle E_I, d_I, \{f_I | f \in F\}\rangle$ such that:

1)$\langle E_I, \{f_I | f \in F\}\rangle$ is an F-algebra,

2)$\langle E_I, d_I\rangle$ is a complete metric space with respect to the distance d_I,

3) for each $f \in F, f_I : E_I^{\rho(f)} \to E_I$, where $\rho(f) \in \mathcal{N}$ is the arity of f, is a continuous function w.r.t. the topology induced by the distance d_I.

A *computation* from $t \in T(F \cup \Phi, E_I)$ with a program P, $P = \langle \Sigma, I\rangle$ as a program scheme Σ and an interpretation I, is a finite or infinite sequence $t_0, \ldots, t_i, \ldots$ of elements of $T(F \cup \Phi, E_I)$ such that $t_0 = t$ and for every $i \geq 0$, $t_i \to_\Sigma t_{i+1}$ or $t_i \to_I t_{i+1}$, where, informaly speaking, $\to_\Sigma$ is a purely syntactic rewriting rule and $\to_I$ is a rewriting rule of the following kind $t \to_I t'$ iff $t = f(e_1, \ldots, e_n)$ with $f \in F, \rho(f) = n$, $e_1, \ldots, e_n \in E_I$ and $t' = f_I(e_1, \ldots, e_n) \in E_I$.

The set of values associated to an element $t \in T(F \cup \Phi, E_I)$ is defined by the mapping $\Pi_I : T(F \cup \Phi, E_I) \to \mathcal{P}(E_I)$, the powerset of (E_I), as follows:

if $t \in E_I$ then $\Pi_I(t) = \{t\}$
if $t \in F_0 = \{f \mid \rho(f) = 0\}$ then $\Pi_I(t) = \{t_I\}$
if $t = f(t_1, \ldots, t_m)$ then $\Pi_I(t) = \{f_I(e_1, \ldots, e_m) \mid e_i \in \Pi_I(t_i)\}$
if $t = \Phi_i(t_1, \ldots, t_{n_i})$ then $\Pi_I(t) = E_I$.

It can be proved that:

1) $t \to_\Sigma t'$ implies $\Pi_I(t) \supseteq \Pi_I(t')$, and
2) $t \to_I t'$ implies $\Pi_I(t) = \Pi_I(t')$

Thus for any computation $t_0, t_1, \ldots$ the sequence $\{\Pi_I(t_n)\}_n$ is decreasing with respect to inclusion.

Denoting by $\delta(\Pi_I(t))$ the diameter of $\Pi_I(t)$, i.e.

$$\delta(\Pi_I(t)) = max\{d_I(e, e') | e, e' \in \Pi_I(t)\}$$

we define, a computation $c = (t_0, t_1, \ldots)$ to be *successful* if $lim_{n \to \infty} \delta(\Pi_I(t_n)) = 0$. The *result* of a successfull computation $c = (t_0, t_1, \ldots)$ is defined to be the singleton set $\cap_n cl(\Pi_I(t_n))$, where $cl(A)$ denotes the topological closure of $A \subset E_I$, in E_I.

Notice that, since E_I is a complete space, the infinite intersection $\cap_n cl(\Pi_I(t_n))$ is in fact a singleton. We shall denote by $Val_{\Sigma,I}(t)$ its element.

2.2. Symbolic-algebraic computations and algebraic semantics

Algebraic semantics consists in specifying as complete metric space that of infinite terms (or trees); the function computed by a program scheme is characterized by a purely formal object: an infinite tree of all formal computations from a term t.

Algebraic-symbolic computations can be viewed in this framework since the set of function symbols on which program schemes are defined correspond to the signature of data types considered in Computer Algebra [A-M79, BER82,JEN79,LOO74] and unknown function symbols correspond to algorithms on those data types.

The basic idea is to introduce a distance between terms, or finite trees, in $T(F, X)$ as follows.

The *depth* on $T(F, X)$ is a mapping $|.| : T(F, X) \to \mathcal{N}$ defined by:

$$|t| = \begin{cases} 1 & \text{if } t \in F_0 \cup X \\ 1 + max\{|t_i|,\ 1 \le i \le n\} & \text{if } t = f(t_1, \ldots, t_n) \end{cases}$$

The *truncation at depth n* of a tree t is defined by the mapping

$$\alpha_n : T(F, X) \to T(F, X \cup \{\Omega\})$$

where $\Omega \notin F \cup X$ with arity 0 which indicates that a branch of a tree has been cut off in the truncation,

$$\alpha_0(t) = \Omega, \quad \text{for every } t$$

$$\alpha_{n+1}(t) = \begin{cases} t & \text{if } t \in F_0 \cup X \\ f(\alpha_n(t_1), \ldots, \alpha_n(t_p)) & \text{if } t = f(t_1, \ldots, t_p) \end{cases}$$

The *distance* between finite trees can now be defined as $d: T(F,X) \times T(F,X) \to \mathcal{R}^+$

$$d(t,t') = \begin{cases} 0 & \text{if } t = t' \\ 2^{-\overline{\alpha}(t,t')} & \text{if } t \neq t' \text{ and } \overline{\alpha}(t,t') = min\{n | \alpha_n(t) \neq \alpha_n(t')\} \end{cases}$$

It can be verified that this distance is an ultrametric one; let us observe that it is essentially the same distance defined on formal series.

The metric space $\langle T(F,X), d\rangle$ is embedded in a complete metric space, noted $T^\infty(F,X)$. The elements of $T^\infty(F,X) - T(F,X)$ are infinite trees.

A *formal computation* from a tree $t \in T(F \cup \Phi, X)$ with a scheme Σ is a finite or infinite sequence $t_0, t_1, \ldots, t_n, \ldots$ of elements of $T(F \cup \Phi, X)$ such that $t_0 = t$ and for every $i \geq 0$, $t_i \to_\Sigma t_{i+1}$.

The set of infinite trees in $T^\infty(F,X)$ associated to a term t in $T(F \cup \Phi, X)$ is defined by a mapping Π,

$$\Pi : T(F \cup \Phi, X) \to \mathcal{P}(T^\infty(F,X))$$

see [A-N80] for the definition.

It can be proved that: $t \to_\Sigma t'$ implies $\Pi(t) \supseteq \Pi(t')$. A formal computation $c = (t_0, t_1, \ldots)$ is said to be *successful* if $lim_{n\to\infty} \delta(\Pi(t_n)) = 0$, where $\delta(\Pi(t))$ is the diameter of $\Pi(t)$.

The *result* of a successful formal computation is the singleton set $\cap_n \Pi(t_n)$, the element of which is denoted by $Val_\Sigma(t)$.

2.3. Integration of computations as equivalence of semantics

One of the main result in the area of program schemes semantics is the equivalence of operational semantics and algebraic semantics. This equivalence can be expressed by the following

Theorem. The result of a successfull computation from $t \in T(F \cup \Phi, E_I)$ with a program $P = \langle \Sigma, I \rangle$ is the interpretation from $t \in T(F \cup \Phi, X)$ with the program scheme Σ, i.e. $Val_{\Sigma,I}(t) = (Val_{\Sigma}(t))_I$. For the proof see [A-N80].

$$
\begin{array}{ccc}
T^{\infty}(F, X) & \longrightarrow & E_I \\
\downarrow & & \downarrow \\
Val_{\Sigma}(t) & \longrightarrow & (Val_{\Sigma}(t)_I) = Val_{\Sigma,I}(t)
\end{array}
$$

This diagram essentially means that it is equivalent to:

1) first compute symbolically obtaining $Val_{\Sigma}(t)$ and then interpret this result obtaining $(Val_{\Sigma}(t)_I)$ and

2) first interpret in a complete metric algebra and then compute in this interpretation obtaining $Val_{\Sigma,I}(t)$.

Since we have formulated the computations on numbers and symbols in terms of operational and algebraic semantics respectively, the integration problem can be expressed by requiring the *commutativity of the diagram.*

However some remarks must be presented here:

(i) the described diagram involves those computations where the numeric (algebraic) steps are carried out only after the execution of all the algebraic (numeric) steps;

(ii) therefore a more accurate formulation of the problem, where numeric and algebraic steps are effectively interleaving, would use the principal of the so called mixed computation, as described in [ERS82];

(iii)the commutativity of a diagram describing the semantic interpretation of this interleaving process must be proved. The approach of semi-interpretation, presented in [C-L83], suggests us how to proceed in that direction.

3. Algebraic and Numeric Computations: Computational Aspects

In this section we look at the integration problem from a computational point of view. We define a computational framework in which the equivalence introduced in the previous sections can be verified and computations on both numbers and functions can be carried out with unique formulations, even by approximation methods, as required in numeric context.

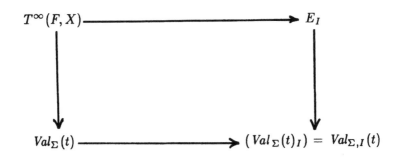

In this paragraph we consider p-adic arithmetic from a computational point of view which needs some further development as pointed out also in [S-V86] in the context of the theory of domains in semantics of programming languages.

We shall now consider a complete metric interpretation of program schemes, namely the complete metric field of p-adic numbers.

The p-adic arithmetic has been introduced by Hensel [HEN08] and for a recent presentation we refer to [KOB77].

Let us consider as particular complete metric F-algebra, where the alphabet is $F = \{+, \star\}$, the complete metric field of p-adic numbers $(Q_p, \tilde{d}_p, +, *, ^{-1})$. The complete space $(Q_p, \tilde{d}_p)$ is defined as the completion of the space (Q_p, d_p) of rational numbers with the p-adic distance $d_p: Q \times Q \to \mathcal{R}^+$,

$$d_p(x,y) = \begin{cases} 0, & \text{if } x = y \\ p^{ord_p(x-y)} & \text{if } x \neq y \end{cases}$$

where

$$ord_p(z) = \begin{cases} \max\{m \in \mathcal{N} \mid z \equiv 0 (mod\ p^m)\} & \text{if } z \in Z \\ ord_p(a) - ord_p(b) & \text{if } z = \frac{a}{b} \in Q - Z \end{cases}$$

It can be verified that d_p is an ultrametric distance, like the distance on trees, i.e. $d_p(x,y) \leq \max(d_p(x,z), d_p(z,y))$.

Notice that $ord_p(x)$ represents the best approximation of x in terms of powers of the prime number p.

As it is known, Q_p is defined as the set of equivalence classes of Cauchy sequences, $\tilde{d}_p$ is defined from d_p and with addition, multiplication and inverses, Q_p is a field.

An important consequence of the ultrametric property of p-adic norm concerns the convergence of p-adic infinite series, expressed by the following

Fact 1. Let $\{c_i\}: \mathcal{N} \to Q_p$ be a sequence of p-adic numbers, the sequence of partial sums $\{\sum_{i=1}^{n} c_i\}$ converges to a limit denoted by $\sum_{i=1}^{\infty} c_i$, if and only if $lim_{i \to \infty} |c_i|_p = 0$.

A p-adic number can bé expressed as the sum of an infinite series, according to the following

Fact 2. Every equivalence class $a \in Q_p$ such that $|a|_p \leq 1$ has exactly one representative Cauchy sequence $\{a_i\}$ such that

$$\left\{ \begin{array}{ll} (1) & 0 \leq a_i < p^i \\ (2) & a_i \equiv a_{i+1} (mod\ p^i) \end{array} \right\} \qquad \text{for } i = 1, 2, \ldots$$

So each p-adic integer $a \in Q_p$, such that $|a|_p \leq 1$, is the sum of the infinite series $\sum_{i=0}^{\infty} b_i p^i$.

Indeed by (1) $a_i = b_0 + b_1 p + \ldots + b_{i-1} p^{i-1}$ with $b_j \in \{0, 1, \ldots, p-1\}$ and by (2) $a_{i+1} = a_i + b_i p^i$, for $i = 1, 2, \ldots$.

Moreover, each p-adic number $a \in Q_p$ such that $|a|_p > 1$, is the infinite series

$$\sum_{i=0}^{m-1} \frac{b_i}{p^{m-i}} + \sum_{i=0}^{\infty} b_{m+i} p^i$$

Indeed, since for such a, there exists $m \in \mathcal{N}$ such that $|ap^m|_p \leq 1$, we have

$$ap^m = \sum_{i=0}^{\infty} b_i p^i = \sum_{i=0}^{m-1} b_i p^i + \sum_{i=0}^{\infty} b_{m+i} p^{m+i}$$

and thus we obtain the above mentioned series for a.

Notice that the uniqueness of the representation asserted by Fact 2, due to the ultrametric property, avoids to have situations like in $\mathcal{R}$, where terminating decimals can also be represented by decimals with repeating 9*s* (i.e. $1 = 0.999\ldots$).

At the same time the p-adic representation has been analysed as a more general method to approximate infinite series functions by sequences of finite series. In fact an arbitrary power series in x over a coefficient domain D can be approximated by arbitrary number of terms, say n, in the domain of finite formal p-adic series $D[[x]]$, over the same coefficient domain D, using the congruence $mod\ x^m$.

Therefore the formal p-adic series introduced before can easily be interpreted as formal p-adic series over an arbitrary abstract domain D, with similar properties as before. In this sense we can consider a unique computational framework with the required completion properties necessary to treat the integration problem. In particular the following results strength this approach.

The p-adic representation of rational numbers and the related arithmetic has been considered by many authors in the recent past. In particular Gregory and Krishnamurthy [G-K84], have also proposed a finite length p-adic representation for a subset of rational numbers.

This very interesting proposal has been recently completed with computationally significant results. Algorithms for both direct and inverse mapping of rational numbers into p-adic representation have been devised [K-G83,MIO84].

Moreover an analysis of the problem of approximating numbers [MIO83] and functions [BGY80,G-Y79,McS78] has brought to a unified view which has a strong common algebraic base. Again the similarity of integers and polynomials has been verified together with the powerful use of algebraic approach to classical numeric problems (i.e. the use of the extended Euclid's algorithm) [MIO83].

These recent results happen to join the well known results about algebraic algorithms for p-adic iteration [YUN76] and the relations between Newton and Hensel constructions [YUN75,LIP76], and all together allow us to make some final remarks:

(i) The unification of numeric and algebraic constructions has been verified for the design and analysis of algorithms following the p-adic approach;

(ii)The properties of p-adic arithmetic suggest interesting similarities between numbers and trees and moreover furnish a solution to the integration problem as expressed by the above mentioned diagram;

(iii)The p-adic arithmetic brings further similarities with polynomials and formal power series, which is not the case for floating point arithmetic;

(iv)The p-adic representation (via the Hensel's Lemma, considered as the algebraic version of the iteration process over the reals) allows also to deal with approximate methods in a unified form.

References

[**ALB80**] R. ALBRECHT: *Roundings and Approximations in Ordered Sets*, Computing Suppl. 2, 17-31 (1980).

[**ADJ77**] ADJ: *Initial algebra semantics and continuous algebras*, JACM 24 68-95 (1977).

[**A-M79**] G. AUSIELLO, G.F. MASCARI: *On the Design of Algebraic Data Structures with the Approach of Abstract Data Types*, Proc. Of EUROSAM 79, Springer Verlag (1979).

[**A-N80**] A. ARNOLD, M. NIVAT: *Metric interpretations of infinite trees and semantics of non deterministic programs*, Theoretical Computer Science 11 181-205 (1980).

[**BCR86**] H.J. BOEHM, R. CARTWRIGHT, M. RIGGLE, O'DONNELL: *Exact Real Arithmetic: A Case Study in Higher Order Programming.* Proc. Lisp Conference (1986).

[**BDG84**] S.R. BALZAC, and others: *The SCRATCHPAD II*, IBM Research Report (1984).

[**BER82**] M. BERGMAN: *Algebraic specifications: a constructive methodology in logic programming*, in Proc. EUROCAM '82, L.N.C.S. vol. 144, Springer Verlag (1982).

[**BGY80**] R.P. BRENT, F.G. GUSTAVSON, D.Y.Y. YUN: *Fast solution of Toeplitz systems of equations and computation of Pade approximants*, J. of Algorithms, 1, (1980).

[**BRO81**] W.S. BROWN: *A simple but Realistic Model of Floating-Point Computation*, ACM Transactions of Mathematical Software, Vol. 7, n. 4 (1981).

[**B-H79**] W.S. BROWN, A.C. HEARN: *Applications of Symbolic Algebraic Computation.* Comp. Comm. 17 (1979).

[**C-L83**] B.COURCELLE, F. LAVANDIER: *A class of program schemes based on tree rewriting systems*, Proc. of CAAP - 83, LNCS n. , L'Aquila (1983).

[**C-N76**] B. COURCELLE, M. NIVAT: *Algebraic families of interpretations*, 17th FOCS 137-146 (1976).

[**C-V85**] M.M. CERIMELE, M. VENTURINI ZILLI: *Effective Numerical Approximations by Intervals*, Freiburger Interwall-Berichte 85/4, 1-24 (1985).

[**E-M85**] E. ENGELER, R. MÄDER: *Scientific Computation: The Integration of Symbolic, Numeric and Graphic Computation*, L.N.C.S. 203 (1985).

[**ERS82**] A.P. ERSHOV: *Mixed computation: Potential applications and problems for study*, Theoretical Computer Sciernce 18, 41-67 (1982).

[**G-K84**] R.T. GREGORY, E.V. KRISHNAMURTHY: *Methods and Applications of Error-Free Computation*, Springer Verlag (1984).

[**G-Y79**] F.G. GUSTAVSON, D.Y.Y. YUN: *Fast algorithms for rational Hermite approximation and solution of Toeplitz systems*, IEEE Trans. Circuits and Systems, CAS 26, n. 9 (1979).

[**HEN08**] K. HENSEL: *Theorie der Algebraischen Zahlen*, Teubner, Leipzig-Stuttgart (1908).

[**JEN79**] R. JENKS: *MODLISP: An Introduction*, Proc. of EUROSAM '79 Conference, Marseille, June 1979, Ed. by Ed, Ng, LNCS, n. 72, 1979.

[**KEM81**] P. KEMP: *Symbolic-numeric interface (abstract)*, SIGSAM Bulletin 15, 2, 6, ACM (1981).

[**KOB77**] N. KOBLITZ: *p-adic Numbers, p-adic Analysis and Zeta Functions*, Springer Verlag (1977).

[**K-G83**] P.KORNERUP, R.T. GREGORY: *Mapping integers and Hensel-codes onto Farey fractions*, DAIMI PB 149, Comp. Sc. Dept. of Aarhus University, Denmark (1982). BIT 23, 9-20 (1983).

[**K-M80**] U. KULISCH, W,L, MIRANKER: *Computer arithmetic in theory and practice*, Academic Press (1980).

[**LIP76**] J.D. LIPSON: *Newton's Method: A Great Algebraic Algorithm*, Proceedings of the 1976 ACM Symposium on Symbolic and Algebraic Computation (1976).

[**LOO74**] R. LOOS: *Toward a Formal Implementation of Computer Algebra*, Proc. of EUROSAM '74, ACM SIGSAM Bulletin, Vol., 8, n. 3 (1974).

[**McS78**] R,J, McELIECE, J.B. SKEARER: *A property of Euclid's algorithm and an application to Pade approximation*, SIAM J. Appl. Math. 34, n. 4 (1978).

[**MIO83**] A. MIOLA: *A unified view of approximate rational arithmetics and rational interpolation*, Proc. of IEEE ARITH-6 Conference, Aarhus (1983).

[**MIO84**] A. MIOLA: *Algebraic Approach to p-adic conversion of Rational Numbers*, IPL 18 (1984).

[**M-T85**] W.L. MIRANKER, R.A. TOUPIN (Eds.): *Accurate Scientific Computations*. L.N.C.S. 235 (1985).

[**M-Y74**] A. MIOLA, D.Y.Y. YUN: *Computational aspects of univariate polynomial GCD*, Proc. ACM EUROSAM '74, SIGSAM Bulletin (8), (1974).

[**NGE79**] E.W. NG: *Symbolic-Numeric Interface: a Review*, Proc. of EUROSAM '79, Springer Verlag (1979).

[**S-V86**] M. SMYTH, S. VICKERS: *The Domain of p-adic Integers*. Report of the 2nd British Theoretical Computer Science Colloquium, Bulletin of the EATCS Number 30, (1986).

[**YUN75**] D.Y.Y. YUN: *Hensel meets Newton-Algebraic constructions in an analytic setting*, IBM Tech. Report RC 5538 (1975).

[**YUN76**] D.Y.Y. YUN: *Algebraic Algorithms using p-adic Constructions*, Proc. 1976 ACM, Symposium on Symbolic and Algebraic Computation (1976).

A SYMBOLIC MANIPULATION SYSTEM FOR COMBINATORIAL PROBLEMS

P. Massazza - G. Mauri
P. Righi - M. Torelli
Dip. Scienze dell'Informazione
Università di Milano - Italy

1. INTRODUCTION

Problems concerning the counting of combinatorial structures are a class of ever growing importance (for example, counting structures is often required in the complexity analysis of algorithms), and it would be therefore very useful to have automatic tools for solving them.

In this paper, a technique for solving counting problems on a rather large class of combinatorial structures is suggested, and a computer algebra system implementing this technique is presented.

As a first step, we have to give a description of combinatorial structures and counting problems over them, using a specification language, for which syntax and semantics are defined.

The description formalism is chosen in such a way that, starting from the given description, a solution of the counting problem in terms of generating functions can be obtained in a natural way.

In fact, the given description can be easily reduced to a system of equations in generating functions, which can be solved mechanically (under suitable conditions which guarantee the existence and uniqueness of the solution).

An interactive system implementing this technique has been realized as an extension of the muMATH System; in Section 4 the architecture and main features of the system are described. As a conclusion, we will briefly discuss the limitations of our approach and the possibility of extending it to counting problems involving subobjects of a given combinatorial object.

2. BASIC DEFINITIONS

Let S be a set of combinatorial objects (typical examples are permutations of a given set, words over a given alphabet and so on). In the following, we will call S a *structure* and its elements *configurations*. Informally, a counting problem consists of determining the cardinality of the subset of all the elements of S having a given property. More formally, one can state:

Def.2.1 - A *weight function* over S is a function $w: S \to N$, where N is the set of natural numbers.

Def.2.2 - Given a weight function w over S, the *counting problem* induced by w (briefly, cp(S,w)) is the problem of computing the function $F: N \to N$ defined by:

$$F(n)=\#\{s\in S \mid w(s)=n\},$$

where #A denotes the number of elements of set A.

The concept of generating function has revealed itself as a general and powerful tool for solving counting problems [10].

Def.2.3 - Given a set S of configurations and a weight function w, the (ordinary) *generating function* of S with respect to w is the function

$$\phi_S^{(w)}(x) = \Sigma_{s\in S}\, x^{w(s)} \;.$$

Now, if one knows the generating function $\phi_S^{(w)}(x)$ of the set S w. r. t. w, the cp(S,w) is solved (on input n) by computing the coefficient a_n in the Taylor series expansion of $\phi_S^{(w)}(x)$. In fact we have:

$$\phi_S^{(w)}(x) = \Sigma_{s\in S}\, x^{w(s)} = \Sigma_{n\geq 0}\, a_n x^n \;\text{ and }\; a_n= \#\{s\in S \mid w(s)=n\}.$$

3. DECOMPOSITION OF STRUCTURES

In many cases, a cp(S,w) can be solved by reduction to a cp(T,w') s.t. there is a weight-preserving isomorphism between S and T and the generating function of T is easier to construct than the one of S [4]. Here, we will treat the particular case where T is obtained from a set of structures (possibly including S), by application of the operations defined below.

Def.3.1 - The *disjoint union* $\biguplus$ of structures is the restriction of the union operation to disjoint structures.

Def.3.2 - The *unambiguous product* $\odot$ of structures is the restriction of the product (concatenation) operation $\cdot$ to structures S_1, S_2 s.t. :

$$\forall\, s \in S_1\odot S_2\ \exists!\ s_1 \in S_1,\, s_2 \in S_2,\, s=s_1\cdot s_2 \,.$$

The weight-preserving isomorphism between S and $T \equiv P(B_1, \ldots, B_n)$, where the operator P uses only the operations $\biguplus$ and $\odot$, will be called a *decomposition* of S. In other words, we need to define an isomorphism $\mathbf{S} \approx \mathbf{P}(\mathbf{S})$, where **S** denotes a vector of structures.

Example 3.1

Consider the set $S = \{0,1\}^*$: we naturally have the isomorphism

$S \approx T \odot S \biguplus V \odot S \biguplus W$

$T \approx \{0\}$

$V \approx \{1\}$

$W \approx \{\varepsilon\}$ (ε is the empty string). ¶

From the decomposition $\mathbf{S} \approx \mathbf{P}(\mathbf{S})$ it is possible to obtain a system of equations having the vector of generating functions of **S** as unknown.

In fact, for every S in **S**, we can operate the substitutions $S \to \phi_S^{(w)}(x)$, $\biguplus \to +$ and $\odot \to \times$, thus obtaining a system of equations of the form:

$$\phi_{\mathbf{S}}^{(w)}(x) = \mathbf{P}'(\phi_{\mathbf{S}}^{(w)}(x)),$$

where **P'** is an operator mapping the semiring of vectors of generating functions into itself. A few combinatorial lemmas guarantee that the generating function of the disjoint union (unambiguos product) of structures S_1 and S_2 is indeed the sum (product) of the generating functions of S_1 and S_2.

Example 3.2

Referring to the previous example we have:

$\phi_S^{(w)}(x) = \phi_T^{(w)}(x) \times \phi_S^{(w)}(x) + \phi_V^{(w)}(x) \times \phi_S^{(w)}(x) + \phi_W^{(w)}(x)$

$\phi_T^{(w)}(x) = \phi_V^{(w)}(x) = x$

$\phi_W^{(w)}(x) = 1$

having chosen w(s) = length of s. ¶

We remark that the choice of only considering decompositions involving the two operators $\biguplus$ and $\odot$ is due to the fact that in this case **P'** results to be a contractive operator over the complete metric space of vectors of generating functions (with a suitable notion of distance). Hence, by Banach's theorem, there is one and only one solution to the system of equations, i.e. the fixed point of the operator **P'**.

4. IMPLEMENTATION

The scheme of solution outlined above has been implemented into a computer algebra system extending the muMATH system [8]. The system has the overall architecture sketched in the following diagram:

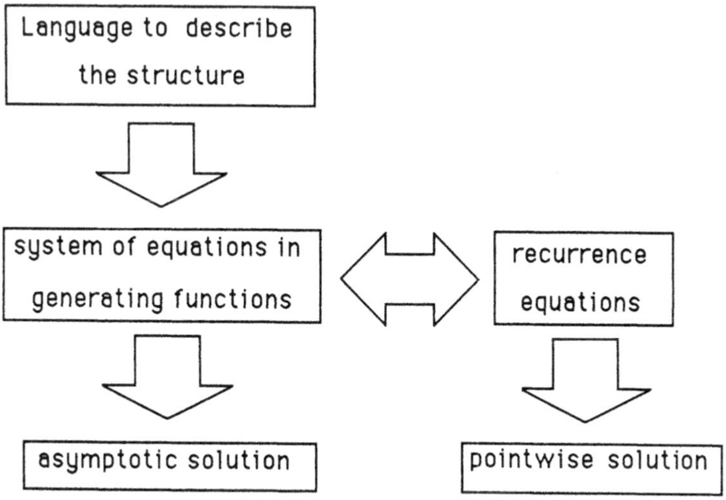

Let us now briefly explain the tasks performed by each module.

4.1. Syntactic and semantic correctness

The module referring to the description of the structure provides a (context free) specification language: it permits the implicit definition of a structure by a decomposition.

Example 4.1.1

Consider the set of strings over the alphabet {a,b} having no adjacent a's: according to the specification language one can write

$$S = b \times S + a \times b \times S + a + e.$$

The interpretation is quite obvious, i. e. "=" stands for "≈", "×" for "⊙", "+" for "⊎", capital letters represent structures, small letters terminal symbols ("e" standing for the empty string). ¶

The system checks the syntactic correctness of the description supplied by the user; more precisely, it parses the input and gives a diagnostic message whenever a syntax error is found. Then, in order to carry out a first semantic check, it classifies the input according to the following three classes: algebraic systems (1), linear systems (2), right (left) linear systems (3). In case (3), starting from the description, a graph <V,E> is constructed, where the set of vertices V is the set of structures, and the set of (labelled) arcs E is the following:

$E \equiv \{(v_i,v_j,\alpha) \mid v_i,v_j \in V$ and $\alpha \times v_j$ is a term of the equation having v_i on the left side$\}$.

This graph is analysed in order to detect ε-labelled loops and vertices which cannot be reached from the vertex representing the structure to be decomposed: these cases correspond to incorrect descriptions. The system, then, proceeds to reduce the right linear description to a regular grammar, in order to detect ambiguity through an efficient algorithm.

The algorithm requires as input a regular grammar $G \equiv \langle N,T,P,S \rangle$, where N is the set of nonterminal symbols, T the set of terminal symbols, S the start symbol (the same symbol indicating the structure under examination), P is the set of productions. Productions are restricted to be of the form:

$A \rightarrow aB$ with $A, B \in N$, $a \in T$ or

$Z \rightarrow \varepsilon$ for a unique symbol $Z \in N$.

Whenever a production $A \rightarrow a$ is encountered in the original grammar, it should be translated into $A \rightarrow aZ$, $Z \rightarrow \varepsilon$.

The algorithm can then be expressed as a function which, on input G, returns "true" if the grammar is ambiguous, "false" otherwise.

```
function ambiguity (<N,T,P,S>);
begin
    w := {(A,B) | ∃ C∈N, a∈T ((C → aA), (C → aB) ∈ P)};
    v := w;
    while (w≠ ∅ and (Z,Z)∉ w) do
    begin
        w := ⋃(A,B) ∈ w, a ∈ T {C∈N | A → aC ∈ P} × {D∈N | B → aD ∈ P};
        w := w \ v;
        v := w ⋃ v;
    end;
    if (w = ∅ )
       then return(false);
       else return(true);
end;
```

Correctness

We distinguish two cases:

1) The function returns the value true, that is at the end of the while-loop it is true that $w \neq \emptyset$, hence $(Z,Z) \in w$. But $(Z,Z) \in w$ iff there exists $(A,B) \in w$ s.t. $A \to aZ$, $B \to aZ$. This means that it is possible to find two derivations for the same string:

$$S \to^* yA \to yaB \to^* yaxZ \to yax \quad \text{and} \quad S \to^* yA \to yaC \to^* yaxZ \to yax$$

so G is ambiguous.

2) The function returns the value false, that is at the end of the while-loop it is true that $w = \emptyset$: this means that the pair (Z,Z) is not reachable from S (otherwise the set w would not be empty, as (Z,Z) would belong to w). Since only Z can produce a string composed by terminal symbols, it follows that there are no distinct derivations of the same string, so G is unambiguous.

Complexity evaluation

The set w of non-terminal pairs is replaced, at every cycle, by a set of non-terminal pairs never reached before. Since the number of such pairs is less than or equal to $|N|^2$, after $O(|N|^2)$ steps the algorithm halts.

If no ambiguity is discovered, then unions and products are guaranteed to be disjoint and unambiguous, respectively.

Since one can prove that for linear systems the ambiguity problem is recursively equivalent to Post correspondence problem (which is known to be undecidable), a warning message is prompted to the user whenever systems of types (1) and (2) are parsed. In these cases, the system cannot automatically ensure the correctness of results, and hence the user must provide an external check with "ad hoc" techniques.

4.2. Pointwise and asymptotic solution

Once the syntactic and semantic analysis has been completed, the description of the structure is automatically translated into a system of equations in generating functions. At this point, the user has to make a choice, i.e. he can obtain the solution of the cp(S,w) for a particular value of n or he can ask for the asymptotic closed form of F(n) (see Section 2).

In the former case a system of recurrence equations is constructed starting from the one in generating functions. In fact, the space of generating functions, viewed as formal power series rather than as analytical functions, is isomorphic to the space of numerical sequences by the following three lemmas: let $\phi_S^{(w)}(x) \approx \{a_n\}$ and $\phi_T^{(w)}(x) \approx \{b_n\}$, then

1) $\phi_S^{(w)}(x) + \phi_T^{(w)}(x) = \phi_U^{(w)}(x) \approx \{c_n\} = \{a_n\} + \{b_n\}$
2) $(1-x)^{-1} \times \phi_S^{(w)}(x) = \phi_U^{(w)}(x) \approx \{c_n\}$ where $c_n = \Sigma_{k=0..n}\, a_k$
3) $x^k \times \phi_S^{(w)}(x) = \phi_U^{(w)}(x) \approx \{c_n\}$ where $c_n =$ if $n \geq k$ then a_{n-k} else 0 .

This construction, although generally applicable, is done by our system only in the linear case. This choice is due to the fact that the resulting recurrence equation system has the form $S_{n+1} = AS_n$, where A is a suitable matrix and S_i a vector of sequences; the solution can be computed efficiently by the formula $S_n = A^n S_o$, using the algebraic manipulation capabilities of the underlying muMATH system.

Example 4.2.1

From example 4.1.1 we have:

$$\phi_S^{(w)}(x) = x \times \phi_S^{(w)}(x) + x^2 \times \phi_S^{(w)}(x) + x + 1$$

from which the system derives the following recurrence equation

$$a_n = a_{n-1} + a_{n-2}$$

and the initial conditions

$$a_0 = 1, \ a_1 = 2.$$

The solution is then automatically computed by the formula

$$a_{n+1} = [1\ 0] \begin{bmatrix} 1 & 1 \\ 1 & 0 \end{bmatrix}^n \begin{bmatrix} 2 \\ 1 \end{bmatrix}$$

¶

In the latter case (closed form of F(n)) the system of equations in generating functions is solved by muMATH and $\phi_S^{(w)}(x)$ is determined in closed form. In the case of a linear system it turns out to be a rational function, i.e. it is of the form $\frac{P(x)}{Q(x)}$ with P(x) and Q(x) polynomials with integer coefficients.

The next step is to analyse the polynomial Q(x) in order to decide whether it has one and only one (real) minimum modulus root x_1, in which case the coefficients of the power series expansion of $\phi_S^{(w)}(x)$ can be asymptotically expressed by the formula $a_n \sim \frac{-c_1}{x_1^{n+1}}$ (for details, cf. [1,2,5]) .In fact (supposing deg(P(x)) < deg(Q(x)) and Q(x) having only simple roots):

$$\frac{P(x)}{Q(x)} = \sum_{i=1}^{m} \frac{c_i}{(x - x_i)} = \sum_{i=1}^{m} \frac{-c_i}{x_i} \sum_{k=0}^{\infty} \left(\frac{x}{x_i}\right)^k$$

where

$$c_i = \lim_{x \to x_i} (x - x_i) \frac{P(x)}{Q(x)} = \frac{P(x_i)}{Q'(x_i)}$$

Therefore

$$\frac{P(x)}{Q(x)} = \sum_{n=0}^{\infty} a_n x^n = \sum_{i=1}^{m} \frac{-c_i}{x_i} \sum_{k=0}^{\infty} \left(\frac{x}{x_i}\right)^k \sim \sum_{k=0}^{\infty} \left(\frac{-c_1}{x_1^{k+1}}\right) x^k , \qquad \text{since } |x_1| < |x_2| \leq \ldots \leq |x_m|.$$

In order to decide whether Q(x) has only one minimum modulus root, the system computes an approximation of this value [3], say $\boldsymbol{x}$, and a lower bound to the distance between roots of Q(x) [7], say S(Q). At this point, the uniqueness of the root is determined by using Sturm sequences [3] on the interval $(\boldsymbol{x} - S(Q) - \varepsilon, \boldsymbol{x} + S(Q) + \varepsilon)$, where ε is a fixed value s.t. $|\mathbf{x} - \boldsymbol{x}| < \varepsilon < S(Q)$, $\mathbf{x}$ being the exact minimum modulus.

Example 4.2.2

In example 4.2.1, muMATH can solve the equation w.r.t. $\phi_S^{(w)}$, obtaining

$$\phi_S^{(w)}(x) = \frac{1 + x}{1 - x - x^2}.$$

The system can then display the asymptotic solution

$$a_n \sim \left(\frac{1}{2} + \frac{\sqrt{5}}{10}\right)\left(\frac{1 + \sqrt{5}}{2}\right)^{n+1}.$$

¶

5. CONCLUSIONS

In the above description, we have stressed some limitations of our system in its current form. A first limitation is the restriction to structures that can be described using only the operations of disjoint union and unambiguous product. In [6], two other operations are considered, i.e. composition and derivation. However, some more work is needed in order to define suitable semantics for a specification language including these operations. A second limitation, i.e. the undecidability of the ambiguity problem in the linear and algebraic case, is of theoretical nature; hence, some results for particular structures can be obtained only by heuristic and non uniform techniques. Presently the system can deal in a completely automatic way with counting problems which admit a description in terms of a regular grammar.

The program consists of 1200 lines of muSIMP code, a LISP-like language in which muMATH is implemented. The choice of using the muMATH system has a twofold motivation:

1) in the muMATH environment it is very simple to add new procedures which use the

capabilities of the existent modules (in a sense muMATH can be regarded as a programming language having a very powerful instruction set)

2) it requires very cheap hardware, in fact all you need for running muMATH is an IBM PC (or compatible).

Of course, muMATH does not possess the full algebraic capabilities of other C.A.S's (MACSYMA, REDUCE, etc.) but, as our system most of all needs elementary linear algebra, this does not constitute a real limitation (that is, by using a "large" C.A.S. you can only obtain a speed-up due to the more powerful hardware, not to the software).

Nevertheless, it should not be too difficult to implement our system on more powerful C.A.S's. (especially if a LISP-like language is available), and this probably will be done when heuristics for the algebraic case will be implemented.

Besides improving the system with respect to the topics quoted above, it would be interesting also to extend it so as to be able to solve counting problems involving the concept of a subobject of a given combinatorial object. Informally, a counting problem with subobjects consists in determining the number of subobjects of a given type in configurations of weight n. A formal definition of subobject is given in [6]. In [9] it is shown how this kind of problems can be specified. However, in this case the associated generating functions are multivariate, and this implies that asymptotic estimations become considerably more complex.

ACKNOWLEDGMENT

The authors gratefully acknowledge the help received from Prof. A. Bertoni, who also inspired this work.

REFERENCES

[1] Bertoni, A., Mauri, G., Torelli, M., Three efficient algorithms for counting problems, *Inform. Proc. Lett.*, 8 (1979), 50-53

[2] Bertoni, A., Mauri, G., Torelli, M., Sulla complessità di alcuni problemi di conteggio, *Calcolo*, 17 (1980), 163-174

[3] Durand, E., *Solutions numérique des équations algébriques*, vol. 1, Masson, 1960

[4] Goulden, I.P., Jackson, D.M., *Combinatorial Enumeration*, Wiley, 1983

[5] Henrici, P., *Applied and Computational Complex Analysis*, vol.1, Wiley, 1974

[6] Massazza, P., *Un sistema di manipolazione simbolica per problemi combinatori: specifiche*

di strutture e soluzioni di problemi di conteggio, Thesis, Dipartimento di Scienze dell'Informazione, Milano, 1985

[7] Mignotte, M., Some Useful Bounds, in *Computer Algebra Symbolic and Algebraic Computation*, B. Buchberger et al. Eds., Springer, 1983

[8] Rich, A.D., Stoutemyer, D.R., Capabilities of the muMATH-79 Computer Algebra System for the Intel 8080 Microprocessor, EUROSAM 1979, 241-248

[9] Righi, P., *Un sistema di manipolazione simbolica per problemi combinatori: specificazione del problema*, Thesis, Dipartimento di Scienze dell'Informazione, Milano, 1985

[10]Stanley, P.R., Generating Functions, in *Studies in Combinatorics* , (G.C. Rota ed.), MAA Studies in Mathematics, vol.17, The Mathematical Association of America, 1978

The present research has been supported by the M. P. I. Project (60%) "Tecniche di manipolazione simbolica".

STANDARD BASES AND NON-NOETHERIANITY: NON-COMMUTATIVE POLYNOMIAL RINGS

Teo Mora

Università di Genova

INTRODUCTION

This paper is a sequel of [MOR1], where I generalized the concept of Groebner bases [BUC1,3] to non-commutative polynomial rings and discussed their main properties and semidecision procedures to compute them.

In the case of commutative polynomial rings over fields and positive term orderings, there are different equivalent characterizations of Groebner bases, the most important (which are just indicated, since concepts are left undefined) being the following:

F is a Groebner basis of an ideal $I \subset K[X_1,\ldots,X_n]$ if:

A: the maximal terms of the basis elements generate the semigroup $M_T(I)$ of all maximal terms of elements in I

B: every $f \in I$ can be represented as $f = \sum g_i f_i$, $f_i \in G$ so that the maximal term of f is not lower than every maximal term of $g_i f_i$

C: the terms which are not in $M_T(I)$ generate a K-vector space which is isomorphic to $K[X_1,\ldots,X_n]/I$, and this isomorphism is computable.

D: property **B** holds for a finite set of polynomials explicitly defined in terms of the basis elements

(remark that **B** and **C** can be expressed in terms of the Buchberger reduction relation and then become respectively: each element in I can be reduced to 0; each element can be reduced to a unique irreducible element).

Properties immediately analogous to these still hold and are equivalent also in the case of non-commutative polynomial rings,so any of them can be chosen as a definition of Groebner bases.

They are, however, no more equivalent in the other known generalizations of Groebner bases; for instance when one considers polynomials over a ring (instead than over a field) [ZAC,SCH,KRK,MOE,PAN] they lead to different concepts of bases.

In particular, when one still works on $P := K[X_1,\ldots,X_n]$, but relaxes the condition of positiveness on the term ordering, so introducing the concept of standard bases [HIR,GAL,LAZ], not only examples suggest that it is difficult to find a statement analogous to **C** [MOR3], but **A** and **B** are equivalent only if stated in some extension R of the polynomial ring P (namely either the localization at the origin or the completion), and then **B** is verified not only by polynomials in I, but also by polynomials in $I R \cap P$, the contraction of the extension of I in R.

The results of Robbiano in his theory of graded structures [ROB1], which is a proposal of a generalization and unification of the known concepts of special bases (Groebner,Macaulay,standard bases) together with other concepts related to the theories of graded and filtered rings, clearly show that in the general case, (the generalizations of) **A** and **B** are not equivalent.

In [MOR1] I gave a definition sufficiently general to allow for a common treatment also of standard bases in non-commutative polynomial rings. This was done choosing **A** as a definition of generalized Groebner bases (which I called d-sets), and using (following [LAZ] and [ROB2]) a concept of term ordering more general than the one introduced by Buchberger [BUC1,3], i.e. relaxing the condition of positiveness for a term ordering.

While it was to be expected that **A** and **B** were no more equivalent for non-positive term orderings, it was possible to give a weaker analogon of **B** (each polynomial in I has a d-representation) which was equivalent to **A**, but which had some unexpected features:

1) the definition doesn't involve representations of polynomials in the ideal through the basis elements, with coefficients chosen in some extension ring;

2) the definition implies that a d-set is not necessarily a basis neither of the ideal (as it already happens for standard bases in the commutative case) nor of its extension in some extension ring.

However, as it will be shown in example 2 below, this analogon of **B** cannot be improved.

The aim of this note is then to give a possible interpretation of its unexpected features: it will come out that the problems are originated more by non-noetherianity than by non-commutativity; and that the analogon of **B** makes perfectly sense, once interpreted in terms of ring completions as follows:

> F is a d-set of I, if, denoting $I^\wedge$ the closure of I in the completion of the polynomial ring w.r.t. topology naturally induced by a term ordering, each element in $I^\wedge$ can be obtained as a limit of a Cauchy sequence of polynomials, each of them with a "good" representation in terms of F.

So a concept, more similar in nature to the one related to representations through Groebner bases, can still be applied, but just to elements "approximating" the elements one has to represent.

This helps to understand the role of the formal power series ring in the definition of (commutative) standard bases, it stresses the fact (already suggested in [ROB1]) that Groebner bases and their generalizations have a strong connections with ring topologies; it suggests also how to state a generalization of **B** which is still equivalent to (the generalization of) **A** in the general context of graded structures.

It can be worthwhile to remark that the results here described for non-commutative polynomial rings were later generalized to the theoretical setting of graded structures [MOR4].

ACKNOWLEDGEMENTS

Comments and remarks by B.Buchberger, A.Logar, L.Robbiano, V.Weispfenning and especially by H.M.Moeller, were helpful to bring this paper into its final form.

1.RECALLS

1.1 If R is a ring and $F \subset R$, we will denote $F^* := \{f \in F : f \neq 0\}$

Let S denote a free semigroup generated by a finite alphabet A.

If m, n are in S, we will say m is a <u>multiple</u> of n (n <u>divides</u> m) iff there are l, r in S s.t. $m = l\,n\,r$.
K[S], K a field, will denote the ring whose elements are finite linear combinations of elements of S:

$$K[S] := \sum_i c_i\, m_i : c_i \in K^*,\ m_i \in S$$

with multiplication canonically defined in terms of the semigroup multiplication.

1.2 A <u>term ordering</u> < on S is a total ordering s.t.:

i) for all m, m_1, m in S, $m_1 < m_2$ implies $m\, m_1 < m\, m_2$ and $m_1\, m < m_2\, m$.

ii) for all m in S, there exists no infinite decreasing sequence $m_1 > \ldots > m_i > \ldots$ s.t. for all i, $m_i > m$.

A term ordering will be called <u>positive</u> iff $1 \le m$ for all m in S; a term ordering will be called <u>negative</u> iff $1 \ge m$ for all m in S or equivalently iff $m \ge m\, n$ and $m \ge n\, m$ for all m, n in S.

1.3 Let < be a term ordering on S. If $f := \sum_{i=1,t} c_i\, m_i$, $c_i \in K^*$, $m_i \in S$, $m_1 > m_2 > \ldots > m_t$, define

$M_T(f) := m_1$, $lc(f) := c_1$.

If $G \subset K[S]$, define $M_T(G) := \{M_T(f) : f \in G^*\}$.
If I is a two-sided non-zero ideal of K[S], $M_T(I)$ is a two-sided ideal of S.

1.4 A <u>distinguished set</u> (shortly a d-set) for I is a set $F \in I^*$ s.t. $M_T(F)$ generates $M_T(I)$.

1.5 We say that f in K[S] has a <u>d-representation</u> in terms of a (possibly infinite) set $F \subset K[S]$ iff there is a sequence $g_1, \ldots, g_i \ldots$ s.t. $g_1 = f$ and for all i:

1) $g_i \in K[S]$
2) if $g_i = 0$ then $g_{i+1} = 0$
3) if $g_i \neq 0$ then there are $l_i, r_i \in S$, $a_i \in K^*$, $f_i \in F$ s.t.:
   i) $g_{i+1} = g_i - a_i\, l_i\, f_i\, r_i$
   ii) $M_T(g_i) = l_i\, M_T(f_i)\, r_i$
   iii) if $g_{i+1} \neq 0$ then $M_T(g_i) > M_T(g_{i+1})$

We say that f has a <u>finite d-representation</u> in terms of F iff:

1) $f = \sum_{i=1,t} a_i\, l_i\, f_i\, r_i$, $a_i \in K^*$, $l_i, r_i \in S$, $f_i \in F$.
2) $M_T(f) = l_1\, M_T(f_1)\, r_1 > l_i\, M_T(f_i)\, r_i \ge l_{i+1}\, M_T(f_{i+1})\, r_{i+1}$ for all $i > 1$

Let us denote, for each m in S, V(m) the K-vector space with basis $\{n : n \in S,\ n < m\}$.
We say that f has a <u>mod. m d-representation</u> in terms of F iff there is g s.t. g has a finite d-representation in terms of F and f - g is in V(m).

1.6 THEOREM The following conditions are equivalent:

1) F is a d-set for I
2) every f in I has a d-representation in terms of F
3) for all f in I, for all m in S, f has a mod. m d-representation in terms of F.

<u>Proof</u>: [MOR1] Prop.2.2.

1.7 Given an ordered pair of terms, $(m_1, m_2) \in S^2$, the <u>set of matches</u> of (m_1, m_2), denoted by $M(m_1, m_2)$, is the finite set of all 4-tuples $(l_1,l_2,r_1,r_2) \in S^4$ s.t. either:

1) $l_1 = r_1 = 1$, $m_1 = l_2\, m_2\, r_2$
2) $l_2 = r_2 = 1$, $m_2 = l_1\, m_1\, r_1$
3) $l_1 = r_2 = 1$, $l_2 \neq 1$, $r_1 \neq 1$, there is $w \in S$ s.t. $w \neq 1$, $m_1 = l_2\, w$, $m_2 = w\, r_1$
4) $l_2 = r_1 = 1$, $l_1 \neq 1$, $r_2 \neq 1$, there is $w \in S$ s.t. $w \neq 1$, $m_1 = w\, r_2$, $m_2 = l_1\, w$.

1.8 LEMMA Let I be an ideal in K[S], $F \subset I^*$ a basis of I, $m \in S$. The following are equivalent:

1) every $g \in I^*$ has a mod. m d-representation in terms of F
2) for all $f_1, f_2 \in F$, for all $(l_1,l_2,r_1,r_2) \in M(m_1,m_2)$, where $m_i := M_T(f_i)$, for all $l, r \in S$,
$f := lc(f_2)\; l\; l_1\, f_1\, r_1\, r - lc(f_1)\; l\; l_2\, f_2\, r_2\, r$
(if not zero) has a mod. m d-representation in terms of F.

Proof: [MOR1] 2.4.

1.9 COROLLARY With the same assumptions as in Lemma 1.8, if < is negative, then the following are equivalent:

1) every $g \in I^*$ has a mod. m d-representation in terms of F
2) for all $f_1, f_2 \in F$, for all $(l_1,l_2,r_1,r_2) \in M(m_1,m_2)$,
$f := lc(f_2)\; l_1\, f_1\, r_1 - lc(f_1)\; l_2\, f_2\, r_2$
(if not zero) has a mod. m d-representation in terms of F.

Proof: 2) $\Rightarrow$ 1) We have just to prove that 1.8.2) holds.
Let then $f_1, f_2 \in F$, $(l_1,l_2,r_1,r_2) \in M(m_1,m_2)$, $f := lc(f_2)\; l_1\, f_1\, r_1 - lc(f_1)\; l_2\, f_2\, r_2$ and assume $f \neq 0$; let $l, r \in S$ and $g := l\, f\, r$.
By assumption we know f has a mod. m d-representation, i.e there is $h \in V(m)$ s.t. $f - h = \sum_{i=1,t} a_i\, l_i\, f_i\, r_i$ is a finite d-representation. Then, since < is negative, $l\, h\, r \in V(m)$ and $g - l\, h\, r = \sum_{i=1,t} a_i\, l\, l_i\, f_i\, r_i\, r$ is a finite d-representation.

2. AN EXAMPLE

2.1. The aim of the following example is to show that, accepting as definition of d-set the one given in 1.5 (i.e. a definition naturally involving "the ideal of maximal terms"), we cannot hope to improve on the concept of d-representation given in 1.5. We will show in fact that it is unavoidable to have not just a "series representation" involving infinitely many summands, but also infinitely many basis elements will be required in the representation.

2.2 Let $A := \{a,b,c,d,e\}$, S the free semigroup generated by A, < any term ordering s.t. for all m,n in S, $deg(m) > deg(n)$ implies $m < n$ (e.g. the inverse of the graduated term ordering defined in [MOR1] 5.11).
Let $f_0 := bedc - cdc^3$, $f_i := ab^ic - ab^{i+2}e$ for $i \geq 1$, so $M_T(f_0) = bedc$, $M_T(f_i) = ab^ic$ if $i \geq 1$.
Because of [MOR1] 2.4, since $M(M_T(f_i),M_T(f_j)) = \emptyset$ for all i,j, $G := \{ f_i : i \in \mathbf{N}\}$ is a d-set for the ideal it generates.

Let $m_i := ab^icdc^{2i-1}$, $i \geq 1$, and $n_i := ab^{i+2}edc^{2i-1}$, $i \geq 1$.

2.3 It is then immediate that the following hold:

1) if $m_i - t = l f_j r$, $t, l, r \in S$, then $j = i$, $t = n_i$.

2) if $t - m_i = l f_j r$, $t, l, r \in S$, then $j = 0$, $i > 1$, $t = n_{i-1}$

3) if $n_i - t = l f_j r$, $t, l, r \in S$, then $j = 0$, $t = m_{i+1}$

4) if $t - n_i = l f_j r$, $t, l, r \in S$, then $j = i$, $t = m_i$

5) $\{t \in S : t - m_1 \in I\} = \{m_i : i \in \mathbf{N}\} \cup \{n_i : i \in \mathbf{N}\}$

6) m_1 is not in I; for all t in S, m_1 is in I + V(t) and has a mod. t d-representation in terms of G: for, let d := deg(t), s s.t. d < 3 s + 2, then

$$m_1 = \sum_{i=1,s-1} f_i\, d\, c^{2i-1} + \sum_{i=1,s-1} a\, b^{i+1} f_0\, c^{2i-2} + m_s$$

is such a representation

7) the only d-representation of m_1 in terms of G is (using with some abuse of notation a "series" representation):

$$m_1 = \sum_{i=1,\infty} f_i\, d\, c^{2i-1} + \sum_{i=1,\infty} a\, b^{i+1} f_0\, c^{2i-2}.$$

2.4. It should then be clear that the concept of d-representation proposed in 1.5 cannot be improved and that no technique as in [MOR2,3] to compute d-representations in finitely many steps can be applied.

3. POLYNOMIALS WITH d-REPRESENTATIONS

3.1 The example above shows another bad feature of d-representations: there are polynomials not in the ideal, which have d-representations in terms of a distinguished set of the ideal.
This parallels a situation which occurs also for standard bases in commutative polynomial rings: there, if I is a polynomial ideal, all polynomials which are in I R, R being either the localization or the completion of the polynomial ring, have a series representation in terms of a standard basis of I.
The aim of this section is to characterize the set of polynomials with d-representations in terms of a d-set of the ideal; we like to give a characterization given in terms of operations within the polynomial ring only; obviously such a characterization is possible also in the commutative case.

3.2 DEFINITION If I is an ideal of K[S], denote $C(I) := \cap_{m \in S} (I + V(m))$.

3.3 REMARK If < is positive, since V(1) = 0, then C(I) = I, so the results below are trivial in this case.
If < is not positive, there is $n \in S$, $n < 1$. There is then an infinite decreasing sequence of terms $n_1, \dots, n_i, \dots$: one obtains such a sequence defining $n_i := n^i$. In the proofs below, we will freely make reference to this sequence.

3.4 PROPOSITION If I is an ideal of K[S], C(I) is an ideal of K[S].
<u>Proof</u>: Remark that if < is positive, C(I) = I, so we need a proof only in the case that there exists $n \in S$,

$n < 1$.

We have to prove that if $f \in C(I)^*$, $g \in K[S]^*$, then $f\,g$ and $g\,f$ are in $I + V(m)$ for every $m \in S$; we will proof just that, given $m \in S$, $f\,g \in I + V(m)$, since the other proof is symmetrical.

Let $n' := M_T(g)$; in the decreasing sequence $n'_1,\ldots,n'_i,\ldots$ where $n'_i := n_i\,n'$, there is i s.t. $n'_i < m$.

Since $f \in I + V(n_i)$, there are $h' \in I$, $h'' \in V(n_i)$, s.t. $f = h' + h''$; therefore $f\,g = h'\,g + h''\,g$, with $h'\,g \in I$ and $h''\,g \in V(n'_i) \subset V(m)$. So $f\,g \in I + V(m)$.

3.5 PROPOSITION The conditions of Theorem 1.6 are equivalent to:

4) $f \in C(I)$ iff f has a d-representation in terms of F.

<u>Proof</u>: 1) $\Rightarrow$ 4): Assume $f \in K[S]^*$ has a d-representation in terms of F. Then (by implication $2 \Rightarrow 3$ of theorem 1.6) for each $m \in S$ there is g s.t. $f - g \in V(m)$ and g has a finite d-representation in terms of F. Then $g \in I$, so $f \in I + V(m)$ for each $m \in S$, and $f \in C(I)$.

Conversely, we want to show that for each $f \in C(I)^*$, f has a d-representation in terms of F. Because of remark 3.3, we have to prove it only in the case $<$ is not positive, and because of Theorem 1.6 we can prove instead that F is a d-set for C(I).

So, let $f \in C(I)^*$, $m := M_T(f)$; since $f \in I + V(m)$, there is $g \in I$ s.t. $f - g \in V(m)$; then $M_T(g) = M_T(f)$, and since F is a d-set for I, $M_T(f)$ is in the ideal generated by $M_T(F)$.

4) $\Rightarrow$ 2): obviously if $f \in I^*$, $f \in C(I)^*$, so it has a d-representation in terms of F.

4. RING COMPLETIONS

4.1 We intend now to give an interpretation of d-representations in terms of ring completions. Therefore, we premise some recalls on basic concepts which will be useful in the following of the paper.

4.2 Let R be an associative ring and, for each $m \in S$, let U(m) be a subgroup of R.

We say $U := \{U(m) : m \in S\}$ is an <u>S-filtration</u> of R if, for each $m, n \in S$, $U(m)\,U(n) \subset U(mn)$.

The S-filtration U induces a topological group structure on R (the one which is obtained considering U as a system of neighborhoods of zero), which is Hausdorff iff $\cap U(m) = 0$.

We say R is an <u>S-filtered ring</u> if, moreover, R is a topological ring w.r.t. the topology induced by U, i.e. if multiplication is continuous.

4.3 If R is an S-filtered ring, with U as filtration, a sequence $(f_i : i \in \mathbf{N})$ of elements of R is called a <u>Cauchy sequence</u> iff for every $m \in S$ there is $n \in \mathbf{N}$ s.t. for all $s, t \geq n$, $f_s - f_t \in U(m)$.

A sequence $(f_i : i \in \mathbf{N})$ <u>converges</u> to $f \in R$ (f is a <u>limit</u> of $(f_i : i \in \mathbf{N})$) iff for every $m \in S$ there is $n \in \mathbf{N}$ s.t. for all $s \geq n$, $f - f_s \in U(m)$.

Two Cauchy sequences (f_i) and (g_i) are called <u>equivalent</u> iff $(f_i - g_i)$ converges to 0.

R is called <u>complete</u> iff each Cauchy sequence of elements of R converges to an element of R.

Each S-filtered ring R has a <u>completion</u> $R\hat{}$ w.r.t. the topology induced by U, i.e. $R\hat{}$ is a topologically complete ring, s.t. R is topologically isomorphic to a dense subring of it (see e.g. [NVO]).

4.4 Clearly, $V := \{V(m) : m \in S\}$ is an S-filtration on K[S], and the topology induced by it is Hausdorff, and moreover is discrete iff $<$ is positive. In this case, K[S] is a complete topological ring with respect to this topology.
Therefore, in the following, we will exclude this trivial case, while, obviously, the results below hold also in this case. So, throughout this section, we will assume $<$ is not positive; $n_1,\ldots,n_i,\ldots$ will denote the infinite decreasing sequence defined in Remark 3.3

4.5 LEMMA K[S] is an S-filtered ring with $V := \{V(m) : m \in S\}$ as S-filtration.
Proof: In order to show that K[S] is an S-filtered ring, we have to prove that multiplication is continuous (cf. 4.2), i.e. that, for each $m \in S$, there are m', m'' in S, s.t. if $f \in V(m')$, $g \in V(m'')$, then $f\,g \in V(m)$.
To prove this, fix an arbitrary m' and let $n'_1,\ldots,n'_i,\ldots$ be the decreasing sequence defined by $n'_i := m'\,n_i$; there is i s.t. $n'_i < m$. Define then $m'' := n_i$.
Then $f\,g \in V(m'\,m'') = V(n'_i) \subset V(m)$.

4.6 We intend here to give a representation of $K[S]^\wedge$ which is different by the one recalled in 4.3.
Define $K[[S,<]]$ to be the set of all applications $f : S \to K$ s.t. there is no infinite increasing sequence (m_i) of elements of S with $f(m_i) \neq 0$ for all i.
$K[[S,<]]$ is given a ring structure, defining:

$(f+g)(m) := f(m) + g(m)$

$(fg)(m) := \sum f(m')\,g(m'')$, where the sum runs on all pairs (which are finitely many) s.t. $m'\,m'' = m$.

Since K[S] can be defined as the ring of those functions $f : S \to K$ which are zero a.e., K[S] can be canonically identified as a subring of $K[[S,<]]$.
This definition naturally extends the definition of the (commutative) "formal power series" ring $K[[X_1,\ldots,X_n]]$, so we will (with the usual abuse of notations) use a "series representation" for the elements of $K[[S,<]]$ which are not in K[S], denoting them as $\sum_{i=1,\infty} c_i\,m_i$, $c_i \in K^*$, $m_i \in S$, $m_i > m_{i+1}$ for all i.
For every such element f, one can define $M_T(f) := m_1$, $lc(f) := c_1$.
One can also define $V(m)^\wedge := \{\, f \in K[[S,<]],\ f = 0 \text{ or } M_T(f) < m \,\}$.
One has then that $V^\wedge := \{V(m)^\wedge : m \in S\}$ is an S-filtration inducing an S-filtered ring structure on $K[[S,<]]$ and that $V(m) = V(m)^\wedge \cap K[S]$.

4.7 LEMMA $K[[S,<]]$ is the completion of K[S].
Proof: 1) $K[[S,<]]$ is complete
Let (f_i) be a Cauchy sequence in $K[[S,<]]$.
We intend to construct a (not necessarily infinite) decreasing sequence $m_1,\ldots,m_i,\ldots$ of elements of S and a sequence $c_1,\ldots,c_i,\ldots$ (indexed on the same set) of elements of K^*, s.t. (f_i) converges to $\sum c_j\,m_j$.
If one can extract from (f_i) an infinite subsequence (g_j) s.t. $M_T(g_j)$ form a decreasing sequence, then (f_i) converges to 0.
Otherwise there is N s.t. if $s \geq N$ then $lc(f_s)\,M_T(f_s)$ is constant.

Define then $m_1 := M_T(f_N)$, $c_1 := lc(f_N)$.
Remark that the Cauchy sequence (g_i) with $g_i := f_i - c_1 m_1$ for all i, is s.t. $M_T(g_i) < m_1$ if i is sufficiently large.
Assume now we have defined $c_1, \ldots, c_n$, $m_1, \ldots, m_n$ s.t. the m_i's are a decreasing sequence and the Cauchy sequence (g_i) with $g_i := f_i - \sum_{j=1,n} c_j m_j$ for all i, is s.t. $M_T(g_i) < m_n$ for sufficiently large i.
Then, again, either one can extract from it an infinite subsequence (h_j) s.t. $M_T(h_j)$ form a decreasing sequence, in which case (g_i) converges to 0, and (f_i) to $\sum_{j=1,n} c_j m_j$; or there is N s.t. $lc(g_s)\, M_T(g_s)$ is constant for $s \geq N$, in which case one defines $m_{n+1} := M_T(f_N)$, $c_{n+1} := lc(f_N)$, and the procedure can be repeated.
If, in this way, one obtains an infinite decreasing sequence $m_1, \ldots, m_j, \ldots$ of elements of S and a corresponding sequence $c_1, \ldots, c_j, \ldots$ of elements of K^*, then clearly (f_i) converges to $g := \sum_{i=1,\infty} c_j m_j$, since for every $m \in S$ there is n s.t. $m_n < m$, and, if s is sufficiently large:

$$M_T(f_s - g) \leq M_T(f_s - \sum_{j=1,n} c_j m_j) < m_n < m.$$

2) <u>For each element</u> f <u>of</u> K[[S,<]] <u>there is a Cauchy sequence in</u> K[S] <u>converging to it</u>
If $f \in K[S]$, then the thesis is obvious. Otherwise let $f =: \sum_{i=1,\infty} c_j m_j$; define $f_n := \sum_{i=1,n} c_j m_j$. Then clearly (f_n) converges to f.

5. RING COMPLETIONS AND d-REPRESENTATIONS

5.1 LEMMA Let $I^\wedge \subset K[[S,<]]$ be the ideal of all limits of Cauchy sequences in I. Then the following hold:

1) $M_T(I^\wedge) = M_T(I)$
2) $I^\wedge = \cap (I^\wedge + V(m)^\wedge)$
3) $C(I) = I^\wedge \cap K[S]$

<u>Proof</u>: 1): Let $f \in I^\wedge$, $f \neq 0$; (f_i) be a Cauchy sequence of elements of I converging to it; by the argument in the proof of Prop.4.7.1), if s is sufficiently large, $M_T(f) = M_T(f_s)$. So the thesis.
2): Let $f \in \cap (I^\wedge + V(m)^\wedge)$; then for each n_i in the decreasing sequence of terms defined in 4.4, there are $f_i \in I^\wedge$, $g_i \in V(n_i)^\wedge$ s.t. $f = f_i + g_i$.
Since f_i is the limit of a Cauchy sequence of elements of I, there is $p_i \in I$ s.t. $f_i - p_i \in V(n_i)^\wedge$. Then (p_i) is a Cauchy sequence of elements of I converging to f, since, for each i, $f - p_i = g_i + (f_i - p_i) \in V(n_i)^\wedge$. Therefore $f \in I^\wedge$.
3): If $f \in I^\wedge \cap K[S]$, then, by the argument above, it is the limit of a Cauchy sequence (p_i) of elements belonging to I. So for each m, if s is sufficiently large, $f - p_s \in V(m)^\wedge \cap K[S] = V(m)$ and, for each m:

$$f = p_s + (f - p_s) \in I + V(m).$$

Therefore $f \in C(I)$.

5.2 LEMMA If $f \in K[S]$ has a d-representation in terms of F, then f is the limit of a Cauchy sequence (p_i) of elements of K[S], s.t. each p_i has a finite d-representation in terms of F.
<u>Proof</u>: Let g_i, a_i, l_i, f_i, r_i be as in 1.5.
For every n, define $p_n := g_1 - g_{n+1} = \sum_{i=1,n} a_i l_i f_i r_i$; then, for every n, p_n has a finite d-representation in

terms of F.

We need to show that (p_n) is a Cauchy sequence converging to f; this is obvious if $g_n = 0$ for large n, so assume $g_n \neq 0$ for every n.

$(M_T(g_n))$ is then a decreasing sequence, so for each $m \in S$ there is n s.t. $M_T(g_n) < m$. Therefore for each $m \in S$, there is n s.t. if $s, t \geq n$:

$$p_t - p_s = g_{s+1} - g_{t+1} \in V(M_T(g_n)) \subset V(m)$$

$$f - p_s = g_{s+1} \in V(M_T(g_n)) \subset V(m).$$

This completes the proof.

5.3 THEOREM The following conditions are equivalent:

1) F is d-set for I

5) $M_T(F)$ generates $M_T(I^\wedge)$

6) $f \in I^\wedge$ iff there is a Cauchy sequence (p_i) of elements of K[S] converging to f, s.t. each p_i has a finite d-representation in terms of F.

Proof: 1) $\Leftrightarrow$ 5) : obvious from Lemma 5.1.1)

6) $\Rightarrow$ 5): Let $m \in M_T(I^\wedge)$, $f \in I^\wedge$ be s.t. $m = M_T(f)$, (p_i) the Cauchy sequence of elements of K[S] converging to f, whose existence is implied by 6).

Then, if s is sufficiently large, $M_T(p_s) = m$, and if $\sum c_i l_i f_i r_i$ is the finite d-representation of p_s in terms of F, $M_T(f) = l_1 M_T(f_1) r_1$.

5) $\Rightarrow$ 6): If f is the limit of a Cauchy sequence (p_i) of elements of K[S], s.t. each p_i has a finite d-representation in terms of F, then for all i, $p_i \in (F) \subset I$, so $f \in I^\wedge$.

Conversely, if $g \in I^\wedge$, there are $f_0 \in F$, $l_0, r_0 \in S$, s.t. $M_T(g) = l_0 M_T(f_0) r_0$.

Then there is $a_0 \in K^*$ s.t. $g_1 := g - a_0 l_0 f_0 r_0$ either is 0 or is s.t. $M_T(g_1) < M_T(g)$.

We can repeat the argument getting a sequence $g = g_0, \ldots, g_n, \ldots$ of elements of $I^\wedge$ s.t., for all i:

1) if $g_i = 0$, then $g_{i+1} = 0$

2) if $g_i \neq 0$, then there are $a_i \in K^*$, $l_i, r_i \in S$, $f_i \in F$, s.t.:

i) $g_{i+1} = g_i - a_i l_i f_i r_i$

ii) $M_T(g_i) = l_i M_T(f_i) r_i$

iii) if $g_{i+1} \neq 0$, then $M_T(g_i) > M_T(g_{i+1})$.

Remark that this is a d-representation except that $g_i \notin K[S]$.

So as in the proof of Prop.5.2, defining $p_n := g_0 - g_{n+1} = \sum_{i=1,n} a_i l_i f_i r_i$, (p_n) is a Cauchy sequence of elements of K[S] converging to f, s.t. each p_i has a finite d-representation in terms of F.

6. TRUNCATED STANDARD BASES

6.1 The interpretation of d-sets provided by Th.5.3 shows that the following definition is a natural one, which extends in a sense the concept of truncated power series. An analogous concept has been recently introduced for the ring of convergent power series in [KFS].

6.2 DEFINITION If $m \in S$, we say F is a m-truncated d-set for an ideal I iff every $f \in I^\wedge$ has a mod. m d-representation in terms of F.

6.3 If $<$ is a negative term ordering, a d-set will be called (as in the commutative case) a standard basis. In this case,every ideal I has a finite m-truncated standard basis for all $m \in S$.
It is obvious that,making use of Lemma 1.8, just minor modifications to Buchberger's algorithm provide an algorithm which,given a finite basis F of a finitely generated ideal I and $m \in S$, computes a finite m-truncated standard basis of I.

7.STANDARD BASES IN COMMUTATIVE POLYNOMIAL RINGS

7.1 Let **X** be a (either finite or enumerable) set of variables $\{X_1,\ldots,X_n,\ldots\}$, and let K[**X**] be the (commutative) polynomial ring in these variables.
The main definitions we have given throughout the paper are generalizations of the analogous definitions for the commutative polynomial ring.
In particular we can define a term ordering $<$ on the commutative free semigroup **T** generated by **X** (whose elements, as usual, we will call terms) as a total ordering s.t.:

i) for all $m, m_1, m_2 \in \mathbf{T}$, $m_1 < m_2$ implies $m\, m_1 < m\, m_2$.

ii) for every $m \in \mathbf{T}$, there exists no infinite decreasing sequence $m_1 > \ldots > m_i > \ldots$ with $m_i > m$ for all i

(remark that this definition doesn't agree either with Buchberger's [BUC1,2],which considers only positive term orderings, nor with Lazard's [LAZ], which doesn't require condition ii)).
We can then define lc(f) and $M_T(f)$ for a polynomial f; $M_T(F)$ for a set F of polynomials, so that $M_T(I)$ is a semigroup ideal, if I is an ideal; V(m) for every term m; $V := \{V(m) : m \in \mathbf{T}\}$; C(I) for every ideal I.
We can introduce also (with just the minor changes required by commutativity) the concepts of d-set, d-representation, finite d-representation, mod. m d-representation.
If we introduce also the concept of **T**-filtered ring, clearly K[**X**] is a **T**-filtered ring and its completion is K[[**X**,<]], whose definition is the commutative analogon of the one given in 4.6.
Remark however that K[[**X**,<]] is just a subring of the formal power series ring K[[**X**]], and coincides with it iff $<$ is negative; otherwise, e.g., if $m > 1$ then $\sum_{i=1,\infty} m^i$ is in K[[**X**]] but not in K[[**X**,<]].
We then have the following analogon of Theorem 1.6, Proposition 3.5 and Theorem 5.3:

7.2 THEOREM The following conditions are equivalent:

1) F is a d-set for I

2) every f in I has a d-representation in terms of F

3) for all f in I, for all term m, f has a mod. m d-representation in terms of F.

4) $f \in C(I)$ iff f has a d-representation in terms of F

5) $M_T(F)$ generates $M_T(I^\wedge)$

6) $f \in I^\wedge$ iff there is a Cauchy sequence (p_i) of elements of K[**X**] converging to f, s.t. each p_i has a finite d-representation in terms of F.

7.3 The following result, however, depends on commutativity; a non commutative counter-example is obtained taking S generated by {a,b}, F := {b - a b a}, I := (F) ⊂ K[S], g := b.

7.4 THEOREM If $M_T(I)$ is finitely generated, (so in particular if K[**X**] is noetherian, i.e. **X** is finite) then the following conditions are equivalent:

1) F is a d-set for I

7) $g \in I^\wedge$ iff $g = \sum_{i=1,t} h_i f_i$, $h_i \in K[[\mathbf{X},<]]$, $f_i \in F$, and $M_T(g) \geq M_T(h_i)\, M_T(f_i)$ for all i.

<u>Proof</u>: 5) ⇒ 7) Since $M_T(I)$ is finitely generated, w.l.o.g. we can assume F is finite.
As in the proof of 5) ⇒ 6) (see 5.3), we can obtain an infinite sequence $g = g_1,\ldots,g_n,\ldots$ of elements of $I^\wedge$ s.t., for all i:

1) if $g_i = 0$, then $g_{i+1} = 0$

2) if $g_i \neq 0$, then there are $a_i \in K^*$, $m_i \in \mathbf{T}$, $f_i \in F$, s.t.:

i) $g_{i+1} = g_i - a_i\, m_i\, f_i$

ii) $M_T(g_i) = m_i\, M_T(f_i)$

iii) if $g_{i+1} \neq 0$, then $M_T(g_i) > M_T(g_{i+1})$

If there is n s.t. $g_n = 0$, then there is nothing to prove.
So, assume $g_n \neq 0$ for all n. For every $f \in F$ define $p_0(f) := 0$.
For all $n \geq 1$, define $p_n(f) := p_{n-1}(f)$ if $f_n \neq f$, $p_n(f) := p_{n-1}(f) + a_n\, m_n$ if $f_n = f$.
Clearly, for every f, ($p_n(f)$) is a Cauchy sequence.
Let p(f) be its limit; then $M_T(p(f))\, M_T(f) \leq m_1\, M_T(f_1) = M_T(g)$. Also, $g = \sum_{f\in F} p(f)\, f$, so the thesis.
7) ⇒ 5) is obvious.

7.5 Also in the commutative case, if $M_T(I)$ is not finitely generated, we cannot improve the concept of d-representation as shown by the following example:
let **X** be infinite; let I ⊂ K[**X**] be the ideal generated by $F := \{f_i : i \geq 1\}$, where $f_i := X_i - X_{i+1}$. Let $w : \mathbf{T} \to \mathbf{N}$ be the unique semigroup morphism s.t. $w(X_i) = i$ and let < be a term ordering s.t. for every m', m" ∈ **T**, w(m') < w(m") implies m' > m", so < is negative and F is a standard basis of I.
Then $X_1 \in I^\wedge$ and X_1 is the limit of the Cauchy sequence (p_n) where for all n, $p_n := X_1 - X_{n+1} = \sum_{i=1,n} f_i$. It is easy to see that this gives the only d-representation of X_1 in terms of F, so F <u>is not a basis of</u> $I^\wedge$ <u>in</u> K[[**X**,<]].
To show that a standard basis F of I may not be a basis of I in K[**X**], also with noetherianity assumptions, the following well-known example can be provided:
Let **X** := {X}, I := (X), $f := X - X^2$, F := {f}, < the unique negative term ordering on **T**, then F is a standard basis of I, but, clearly $X \notin (F)$ and the only representation of X in terms of F satisfying the conditions of Th.7.4.7) is: $X = (\sum_{i=1,\infty} X^i)\, f$.

REFERENCES

[BUC1] B.BUCHBERGER Ein Algorithmus zum Auffinden der Basiselemente des Restklassenringes nach einem nulldimensionalen Polynomideal, Ph.D. Thesis, Innsbruck (1965)

[BUC2] B.BUCHBERGER A criterion for detecting unnecessary reductions in the constructions of Groebner bases, Proc. EUROSAM 79, L. N. Comp. Sci. 72 (1979), 3-21

[BUC3] B.BUCHBERGER Groebner bases: an algorithmic method in polynomial ideal theory, in N.K.BOSE Ed., Recent trends in multidimensional systems theory, Reidel (1985)

[GAL] A.GALLIGO A propos du theoreme de preparation de Weierstrass, L. N. Math. 409 (1974), 543-579

[HIR] H.HIRONAKA Resolution of singularities of an algebraic variety over a field of characteristic zero, Ann. Math. 79 (1964), 109-326

[KRK] A.KANDRI-RODY, D.KAPUR An algorithm for computing the Groebner basis of a polynomial ideal over a euclidean ring (1984)

[KFS] H.KOBAYASHI, A.FURUKAWA, T.SASAKI Groebner basis of ideal of convergent power series (1985)

[LAZ] D.LAZARD Groebner bases, Gaussian elimination, and resolution of systems of algebraic equations, Proc. EUROCAL 83, L. N. Comp. Sci. 162 (1983), 146-156

[MOE] H.M.MOELLER On the construction of Groebner bases using syzygies (1986)

[MOR1] F.MORA Groebner bases for non-commutative polynomial rings, Proc. AAECC3, L. N. Comp. Sci. 228 (1986)

[MOR2] F.MORA An algorithm to compute the equations of tangent cones, Proc. EUROCAM 82, L. N. Comp. Sci. 144 (1982), 158-165

[MOR3] F.MORA A constructive characterization of standard bases, Boll. U.M.I., Sez.D, 2 (1983), 41-50

[MOR4] T.MORA Standard bases (1), Congress on "Computational Geometry and Topology and Computation in Teaching Mathematics" Sevilla (1987)

[NVO] C.NASTASESCU, F. VAN OYSTAEYEN Graded Ring Theory, North-Holland (1982)

[PAN] L.PAN On the D-bases of ideals in polynomial rings over a principal ideal domain (1985)

[ROB1] L.ROBBIANO On the theory of graded structures, J. Symb. Comp. 2 (1986)

[ROB2] L.ROBBIANO Term orderings on the polynomial ring Proc. EUROCAL 85, L. N. Comp. Sci. 204 (1985), 513-517

[SCH] S.SCHALLER Algorithmic aspects of polynomial residue class rings, Ph.D. Thesis, Wisconsin Univ. (1979)

[ZAC] G.ZACHARIAS Generalized Groebner bases in commutative polynomial rings, Bachelor Thesis, M.I.T. (1978)

EMBEDDING SELF-DUAL CODES INTO PROJECTIVE SPACE

H. Opolka
Mathematisches Institut
Universität Göttingen
D - 3400 Göttingen

Abstract: Using the concept of the Weil-representaion we identify self-dual codes as points in a finite dimensional complex projective space.

Let $k = \mathbb{F}_q$ be the finite field with q elements and let k^g be the standard vector space of dimension g over k. Let

$$\langle\, ,\, \rangle : k^g \times k^g \to k\ , \quad \langle x,y\rangle = x_1y_1 + \ldots + x_gy_g\ ,$$

$x = (x_1,\ldots,x_g)$, $y = (y_1,\ldots,y_g) \in k^g$, be the usual scalar product. On $W = k^g \times k^g$ consider the symplectic pairing

$$\omega : W \times W \to k\ , \quad \omega((x,y),(x',y')) = \langle x,y'\rangle - \langle x',y\rangle\ ,$$

$x,y,x',y' \in k^g$. A linear subspace $U \leq W$ is called symplectic selfdual if $U = U^{\perp}$ where $U^{\perp} := \{w \in W \mid \omega(w,u) = 0 \text{ for all } u \in U\}$. Note that every linear subspace $V \leq k^g$ which is selfdual with respect to the usual scalar product, i.e. which satisfies $V = V^*$ where $V^* := \{v \in k^g \mid \langle v,v'\rangle = 0 \text{ for all } v' \in V\}$, gives the symplectic selfdual subspace $V \times V \leq W$. Let $H = H(W)$ be the Heisenberg group corresponding to the pair (W,ω), i.e. $H = \{(w,s) \mid w \in W\ , s \in k\}$ with group law $(w,s)(w',s') = (w+w',\ s+s'+\langle w_1,w_2'\rangle)$, $w = (w_1,w_2)$, $w' = (w_1',w_2') \in W$. Let $\chi : k \to \mathbb{C}^*$ be the standard character, i.e. $\chi(x) = \exp(2\pi i\cdot \mathrm{tr}(x))$ where $\mathrm{tr}: k \to \mathbb{F}_p$ is the trace map of k to its prime field $\mathbb{F}_p$. It is well known that there is up to isomorphism a unique irreducible representation ρ of H with central character χ. Let $Sp = Sp(W,\omega)$ denote the symplectic group of (W,ω) and let $\tilde{S}p$ denote its unique 2-fold central extension. $\tilde{S}p$ acts on H, fixing exactly the center which is isomorphic to k. There is a representation r of $\tilde{S}p$ - which is unique up to isomorphism in "most cases" - on the representation space of ρ such that

$$r(\tilde{g})\rho(h)r(\tilde{g}^{-1}) = \rho(ghg^{-1})\ , \quad \tilde{g} \in \tilde{S}p\ , \ h \in H\ . \tag{1}$$

r is called the Weil-representation corresponding to (W,ω). For any subgroup $A \leq H$ define $A^{\#}$ to be the centralizer of A in H modulo the kernel of χ. If $A = A^{\#}$ we say that A is a polarization. In this case $k \leq A$ and $A = (A \cap W) \oplus k$. The set $A \cap W$ is a subspace of W which is symplectic selfdual. Conversely, if $A \leq H$ is any subgroup such that $A \cap W$ is a symplectic selfdual subspace of W then $A = A^{\#}$, i.e. A is a polarization. Summarizing we have

(2) <u>There is a bijective correspondence between the subgroups</u> $A \leq H$ <u>which are polarizations and subspaces of</u> W <u>which are symplectic selfdual; it is given by</u> $A \to A \cap W$.

Let Ω denote the set of all polarizations. The action of Sp on H permutes the polarizations and Ω consists of $2g+1$ orbits for Sp . Let $A \leq H$ be a polarization. Then there is a unique extension χ_A of the standard character χ of k to A such that $A \cap W \leq \operatorname{Ker}(\chi_A)$. By computing characters we obtain the following proposition.

(3) <u>Let</u> $A \leq H$ <u>be any polarization. Then the irreducible representation</u> ρ <u>of</u> H <u>with central character</u> χ <u>is isomorphic to the representation</u> ρ_A <u>which is induced by</u> χ_A .

The representation ρ_A can be realized on the space S_A of all functions $\phi: H \to \mathbb{C}$ with the property $\phi(h'a) = \chi_A^{-1}(a)\,\phi(h')$, $h' \in H$, $a \in A$, by defining $(\rho(h)(\phi))(h') = \phi(h^{-1}h')$, $h,h' \in H$. A basic observation due to Cartier is as follows.

(4) <u>Let</u> $\mathfrak{F}$ <u>be a representation space for</u> ρ . <u>Then for any polarization</u> $A \leq H$ <u>there is a linear functional</u> $\lambda_A: \mathfrak{F} \to \mathbb{C}$ <u>with the following property</u>

$$\lambda_A(\rho(a)(v)) = \chi_A(a)\lambda_A(v) \ , \ a \in A \ , \ v \in \mathfrak{F} \quad .$$

λ_A <u>is specified up to scalars by this equation.</u>

To define λ_A let $f: \mathfrak{F} \to S_A$ be an isomorphism between ρ and ρ_A . Then $\lambda_A(v) := f(v)(1)$, $v \in \mathfrak{F}$, 1 = unit element in H . Statement (4) gives an embedding of Ω into the projective space of the vector space $\mathfrak{F}^*$ which is dual to $\mathfrak{F}$:

(5) $$\Omega \hookrightarrow \mathbb{P}(\mathfrak{F}^*) \ , \ A \to \lambda_A \ .$$

This embedding is obviously equivariant with respect to the natural action of Sp on Ω and the action of Sp on $\mathbb{P}(\mathfrak{F}^*)$ which is induced by the projective Weil-representation r of Sp defined under (1) . It allows to define the height of a selfdual code as the usual height of the corresponding point in projective space. One can then ask for the properties of a selfdual code which are reflected by its height. All this should be seen in analogy to the projective embedding of modular curves. The mathematical framework for the above point of view is of course well known.

Reference

R. Howe: θ - series and invariant theory, in :
AMS proceedings of symposia in pure math., Vol. 33, 1979, part 1 , 275 - 285

H. Opolka
Mathematisches Institut
Bunsenstr. 3-5

3400 Göttingen

THE FINITE FOURIER-TRANSFORM AS A MODULAR SUBSTITUTION

H. Opolka

Mathematisches Institut
Universität Göttingen
D - 3400 Göttingen

Abstract: We prove that the exact weight enumerator of a linear self-dual code of lenght $g \geq 1$ defines a modular form of weight 1/2 on the Siegel upper half plane H_g for a certain level subgroup of $Sp(2g,\mathbb{Z})$ which contains the $(2g \times 2g)$-matrix $\begin{pmatrix} 0 & 1_g \\ -1_g & 0 \end{pmatrix}$. This result depends on a realization of the finite Fourier-transform as a modular substitution. We explain also a general idea relating linear codes to complex function theory and algebraic geometry.

At first we describe our main result. Let l be a prime, let $g \geq 1$ be an integer and let V be the $\mathbb{F}_l$ - vector space $\mathbb{F}_l^g$. Denote by $S(\mathbb{F}_l)$ the $\mathbb{C}$-vector space of functions $f: \mathbb{F}_l \to \mathbb{C}$ and by $T_g(\mathbb{F}_l)$ the g-fold tensor product of $S(\mathbb{F}_l)$. We view the exact weight enumerator P_C of a linear code $C \leq \mathbb{F}_l^g$ as an element of $T_g(\mathbb{F}_l)$. Let $\mathcal{M}^g$ be the $\mathbb{C}$-vector space of all modular forms of weight 1/2 on the Siegel upper half space H_g.

<u>Theorem</u> There is a $\mathbb{C}$-linear map

$$\mu_l : T_g(\mathbb{F}_l) \to \mathcal{M}^g$$

such that for every linear code $C \leq V$ the image of the exact weight enumerator $P_C \in T_g(\mathbb{F}_l)$ under μ_l is a nontrivial modular form of weight 1/2 on H_g for a certain level subgroup $\Gamma_C \leq Sp(2g,\mathbb{Z})$. There is an explicit linear expression for $\mu_l(P_C)$ in terms of theta functions. If C is self-dual then Γ_C contains the matrix $\begin{pmatrix} 0 & 1_g \\ -1_g & 0 \end{pmatrix}$.

The theory of theta functions suggests relations between linear codes, complex function theory and algebraic geometry. We explain this general idea in more detail: Let $g \geq 1$ be an integer and let H_g denote the so called Siegel upper half plane of degree g, i.e. H_g consists of all complex symmetric $g \times g$-matrices Ω whose imaginary part is positive definite. H_g is considered as an open subset of $\mathbb{C}^{g(g+1)/2}$. The basic theta function is the function on $\mathbb{C}^g \times H_g$ defined by the series

(1) $$\vartheta(z,\Omega) = \sum_{n\in\mathbb{Z}^g} \exp(\pi i\ {}^t n\Omega n + 2\pi i\ {}^t nz) ;$$

here n , z are thought of as column vectors and ${}^t nz$ is the dot product of the row vectors ${}^t n$ and z . It is well known and easy to prove that ϑ converges absolutely and uniformly in z and Ω in each set of the form

$$\max |\mathrm{Im}(z)| < \mathrm{constant}/2\pi , \mathrm{Im}(\Omega) \geq \mathrm{constant} \cdot \mathbf{1}_g ,$$

and therefore represents a holomorphic function in z and Ω . The theta function with rational characteristics $\alpha, \beta \in \mathbb{Q}^g$ is defined by

(2) $$\vartheta[{}^{\alpha}_{\beta}] (z,\Omega) = \exp(\pi i\ {}^t a\Omega a + 2\pi i\ {}^t a(z + b))\ \vartheta(z + \Omega a + b,\Omega) .$$

Fix $\Omega \in H_g$ and let $L := \mathbb{Z} \oplus \Omega\cdot\mathbb{Z}$ be the lattice in $\mathbb{C}^g$ which is generated by the unit vector and by the columns of Ω . For any integer $l \geq 1$ the functions

(3) $$f_a(z,\Omega) := \vartheta[{}^{a/l}_{0}] (lz,l\Omega) , a = (a_1,\ldots,a_g) , 0 \leq a_i < l$$

or

(4) $$g_b(z,\Omega) := \vartheta[{}^{0}_{b/l}] (z,\tfrac{1}{l}\Omega) , b = (b_1,\ldots,b_g) , 0 \leq b_i < l$$

can be used to embed the complex torus $\mathbb{C}^g/L$ holomorphically into a complex projective space:

$$\mathbb{C}^g/L \hookrightarrow \mathbb{P}^{l^g-1} , z \to (\ldots,f_a(z,\Omega),\ldots)_a .$$

Hence by a well known result of algebraic geometry the image is an algebraic variety, and a complete set of defining equations can be obtained from certain theta relations, see e.g. [M] , chapter II . Now fix $z \in \mathbb{C}^g$, e.g. $z = 0$. Then for all $\alpha,\beta \in \mathbb{Q}^g$, $l \in \mathbb{Z}$, $l \geq 1$, the functions $\Omega \to \vartheta[{}^{\alpha}_{\beta}] (0,l\ \Omega)$ are modular forms of weight 1/2 on H_g for a suitable level subgroup $\Gamma \leq Sp(2g, \mathbb{Z})$ (the precise definition will be given below). They can be used to embed certain compactified quotients M_g of H_g holomorphically into a complex projective space:

$$M_g \hookrightarrow \mathbb{P}^{l^g-1} , \Omega \to (\ldots,f_a(0,\Omega),\ldots)_a .$$

The image of this embedding is an algebraic variety and defining equations can again be obtained from certain theta relations, see e.g. [C] .

The subgroup $\Gamma_{1,2}$ of the symplectic group $Sp(2g, \mathbb{Z})$,

$$\Gamma_{1,2} = \{\gamma = \begin{pmatrix} A & B \\ C & D \end{pmatrix} \in Sp(2g, \mathbb{Z}) \quad \begin{matrix} \text{diagonal } {}^tAC \text{ even} \\ \text{diagonal } {}^tBC \text{ even} \end{matrix} \} ,$$

which is generated by the matrices

$$\begin{pmatrix} 0 & 1_g \\ -1_g & 0 \end{pmatrix}, \quad \begin{pmatrix} A & 0 \\ 0 & {}^tA^{-1} \end{pmatrix}, \quad \begin{pmatrix} 1_g & B \\ 0 & 1_g \end{pmatrix} ,$$

$A \in GL(g, \mathbb{Z})$, B symmetric integral with even diagonal, acts on $\mathbb{Q}^g \times \mathbb{Q}^g$, on $\mathbb{C}^g$ and on H_g . All these actions are related by the theta functional equation which will be explained later; for details see [M] , chapter II, § 5 .
So viewing theta characteristics $[{a/l \atop 0}]$, $a = (a_1,\dots,a_g)$, $0 \leq a_i < 1$, as "words" taken form $\frac{1}{l}\ \mathbb{Z}^g/\mathbb{Z}^g$ or from a linear code $C \leq \frac{1}{l}\ \mathbb{Z}^g/\mathbb{Z}^g$ the functional equation leads us to conjecture - rather naively - relations between linear codes, complex function theory and algebraic geometry.
In [O] we had briefly discussed the situation where $\Omega \in H_g$ is fixed. In this note we shall fix z , more presisely we shall take $z = 0$, and allow Ω to vary over H_g . The exact weight enumerator of C is then shown to correspond to the sum of the functions $f_a(0,\Omega)$ on H_g , $(a/l) \bmod \mathbb{Z}^g \in C$, and in the special case where C is selfdual this sum is considered in more detail.

For every finite abelian group A denote by $S(A)$ the $\mathbb{C}$-vector space of all functions $f:A \to \mathbb{C}$. The finite Fourier-transform corresponding to $V = \mathbb{F}_l^g$ is the mapping $F_V : S(V) \to S(V)$ defined by

$$F_V(f)(w) := \sum_{v \in V} \exp(2\pi i(v,w)/l)f(v)$$

where $(v,w) = v_1w_1 + \dots + v_gw_g$, $v = (v_1,\dots,w_g)$, $w = (w_1,\dots,w_g)$. Let E_X denote the characteristic function of a subset $X \subset V$. Clearly, the functions $E_v := E_{\{v\}}$, $v \in V$, form a basis of $S(V)$. It is easily shown that for every linear code $C \leq V$ the following relation holds

$$(5) \qquad F_V(E_C) = |C| \cdot E_{C^\perp} .$$

Under suitable identifications this identity is recognized in [T] as the MacWilliams identity for the exact weight enumerator P_C of C , viewed as an element in $T_g(\mathbb{F}_l)$. Namely, an arbitrary element in $T_g(\mathbb{F}_l)$ can be written uniquely in the form $\Sigma_v a_v x^v$, $v \in V$, with coefficients $a_v \in \mathbb{C}$, where $x^v = x_{v_1} \otimes \dots \otimes x_{v_g}$, $v = (v_1,\dots,v_g)$, $x_{v_j} \in S(\mathbb{F}_l)$, $x_{v_j}(k) = \delta_{jk}$, and the inverse of the $\mathbb{C}$-linear isomorphism

$$\phi = \phi_g : T_g(\mathbb{F}_l) \tilde{\to} S(V) \ , \ x^v \to E_v \ ,$$

transforms the identity (5) into the MacWilliams identity for the exact weight enumerator $P_C(x) = \sum_{v \in C} x^v$:

(6)
$$P_C(y) = |C| \cdot P_{C^\perp}(y) \ ,$$

where $y = (\dots, y_j, \dots)$, $0 \le j < l$, and

$$y_j = \sum_{k=0}^{l-1} \exp(2\pi i k j / l) x_k \ .$$

Now let $g \ge 1$ be an integer and let $H_g \subset \mathbb{C}^{g(g+1)/2}$ denote the Siegel upper half space. There are many relations between the theta functions defined under (1)-(4), for instance the very simple linear relation

$$g_b(z,\Omega) = \sum_a \exp(2\pi i \, {}^t ab/l) f_a(z,\Omega)$$

where a runs over all integral vectors $a = (a_1,\dots,a_g)$, $0 \le a_i < l$. It means that g_b is the finite Fourier-transform of f_a with respect to the characteristics:

(7)
$$g_b = \hat{f}_a \ .$$

Deeper relations are provided by the so called functional equation. It describes the action of the group $\Gamma_{1,2}$ on theta functions with characteristics. We need only the case of "Thetanullwerte" , see e.g. [M] , p. 190 : For all $\Omega \in H_g$; $\alpha_1, \alpha_2 \in \mathbb{Q}^g$; $\gamma = \binom{AB}{CD} \in \Gamma_{1,2}$

(8)
$$\vartheta\begin{bmatrix} D\alpha_1 - C\alpha_2 \\ -B\alpha_1 + A\alpha_2 \end{bmatrix} (0,(A\Omega+B)(C\Omega+D)^{-1}) =$$

$$= \zeta_\gamma \cdot \det(C\Omega+D)^{1/2} \cdot \rho(\gamma,\alpha_1,\alpha_2) \cdot \vartheta\begin{bmatrix}\alpha_1 \\ \alpha_2\end{bmatrix} (0,\Omega)$$

where $\zeta_\gamma^8 = 1$ and ζ_γ depends only on γ , not on α_1, α_2, Ω, and

$$\rho(\gamma,\alpha_1,\alpha_2) = \exp(-\pi i \, {}^t\alpha_1 \, {}^tBD\alpha_1 + 2\pi i \, {}^t\alpha_1 \, {}^tBC\alpha_2 - \pi i \, {}^t\alpha_2 \, {}^tAC\alpha_2) \ .$$

Especially for $\alpha_1 = 0$, $\alpha_2 = a/l$, $a = (a_1,\dots,a_g)$, $0 \le a_i < l$; $\gamma = J = \begin{pmatrix} 0 & 1_g \\ -1_g & 0^g \end{pmatrix} \in \Gamma_{1,2}$ we obtain from (8)

$$\vartheta[{}^{a/l}_{0}]\ (0,(-\Omega)^{-1}) = \xi_j \cdot \det(-\Omega)^{1/2}\ \vartheta[{}^{0}_{a/l}]\ (0,\Omega)$$

or, in view of (7) ,

$$\hat{f}_a(0,\Omega) = \xi_j^{-1} \det\left(-\frac{1}{l}\Omega\right)^{-1/2} \vartheta[{}^{0}_{a/l}]\ \left(0,\left(-\frac{1}{l}\Omega\right)^{-1}\right) .$$

This equation realizes the finite Fourier-transform as a modular substitution.

We need also the concept of a modular form of weight 1/2. Let $\Gamma \leq \Gamma_{1,2}$ be a subgroup of finite index. Then a modular form on H_g of weight $k \in (1/2)\mathbb{Z}$ and level Γ is a holomorphic function f on H_g such that

(10) $$\frac{(f|\gamma)(\Omega)}{(\vartheta|\gamma)[{}^{o}_{o}](0,\Omega)^{2k}} = \frac{f(\Omega)}{\vartheta[{}^{o}_{o}](0,\Omega)} \quad \text{for all } \Omega \in H_g\ ,\ \gamma \in \Gamma .$$

Here we have used the notation

$$(f|\gamma)(\Omega) = f((A\Omega+B)(C\Omega+D)^{-1})\ ,\ \gamma = \begin{pmatrix} A & B \\ C & D \end{pmatrix} \in \Gamma_{1,2} .$$

Recall that for all $\alpha_1,\alpha_2 \in \mathbb{Q}^g$, $l \in \mathbb{Z}$, $l \geq 1$,

$$\vartheta[{}^{\alpha_1}_{\alpha_2}]\ (0,l\cdot\Omega)$$

is a modular form of weight 1/2 for a suitable level Γ which can be computed, see [M], p. 200. Let $\mathcal{M}^g(\Gamma)$ denote the $\mathbb{C}$-vector space of all modular forms on H_g of weight 1/2 and level Γ and let $\mathcal{M}^g$ denote the $\mathbb{C}$-vector space of modular forms on H_g of weight $\frac{1}{2}$ of any level. Let $S(\mathbb{A}_f^g)$ denote the Schwartz-space of $\mathbb{A}_f^g$, $\mathbb{A}_f$ = = finite adeles of $\mathbb{Q}$. It can be identified with the $\mathbb{C}$-vector space generated by all functions $h:\mathbb{Q}^g \to \mathbb{C}$ which are inflated by a function $\frac{1}{m}\mathbb{Z}^g/\mathbb{Z}^g \to \mathbb{C}$ for some integer $m \geq 1$. For every integer $l \geq 1$ there is the so called Weil-map, see [M], p. 2o1,

$$\omega_l : S(\mathbb{A}_f^g) \to \mathcal{M}^g$$

defined by

$$\omega_l(h)(\Omega) := \sum_{n\in\mathbb{Q}^g} h(n)\exp(\pi i\ {}^t n l\Omega n) .$$

It can be shown that its image is the $\mathbb{C}$-vector space generated by the functions $\vartheta[{}^{\alpha_1}_{\alpha_2}]\ (0,l\Omega)$, $\alpha_1,\alpha_2 \in \mathbb{Q}^g$. For l prime consider the composition of maps

$$\mu_1 : T_g(\mathbb{F}_1) \overset{\Phi}{\to} S(V) \overset{i}{\to} S(\mathbb{A}_f^g) \overset{\omega_1}{\to} \mathcal{M}^g$$

where the map i is defined by identifying V with $\tilde{V} = \frac{1}{l}\mathbb{Z}^g/\mathbb{Z}^g$, $\tilde{V} \cong V$, $(a/l) \bmod \mathbb{Z}^g \to v_a$, $v_a = (a_1,\dots,a_g)$, $0 \le a_i < l$, and viewing a function $f : V \to \mathbb{C}$ as a function on $\tilde{V}$ and then by inflation on $\mathbb{A}_f^g$. Under this map the exact weight enumerator $P_C \in T_g(\mathbb{F}_1)$ of a linear code $C \le V$ corresponds to the sum of the functions $f_a(0,\Omega)$, $v_a \in C$:

$$\mu_1(P_C)(\Omega) = \vartheta_C(\Omega) := \sum_{v_a \in C} f_a(0,\Omega) , \tag{11}$$

and the MacWilliams identity (6) becomes

$$\sum_{v_b \in C} g_b(0,\Omega) = |C| \cdot \sum_{v_c \in C^\perp} f_c(0,\Omega) . \tag{12}$$

Now assume $C = C^\perp$. Then the last equation implies

$$\sum_{v_b \in C} g_b(0,\Omega) = |C| \cdot \sum_{v_c \in C} f_c(0,\Omega) . \tag{13}$$

Let $\Gamma \le \Gamma_{1,2}$ be a level for the function $\vartheta_C \in \mathcal{M}^g$. In order to prove the theorem we have to show that Γ contains $J = \begin{pmatrix} 0 & 1_g \\ -1_g & 0 \end{pmatrix} \in Sp(2g, \mathbb{Z})$. This is seen from the following computation:

$$\frac{(\vartheta_C | J(\Omega))}{(\vartheta[{}^o_o] | J(0,\Omega))} = \frac{\vartheta_C(-\Omega^{-1})}{\vartheta[{}^o_o](0,-\Omega^{-1})} = \frac{\sum_{v_a \in C} \vartheta[{}^{a/l}_0]\,(0, l(-\Omega)^{-1})}{\vartheta[{}^o_o]\,(0,-\Omega^{-1})}$$

which by (13) is

$$= \frac{\sum_{v_a \in C} \vartheta[{}^{0}_{a/l}]\,(0, \tfrac{1}{l}(-\Omega)^{-1})}{|C| \cdot \vartheta[{}^o_o]\,(0, -\Omega^{-1})} ,$$

and the relation (9) shows this equal to

$$= \frac{\sum_{v_a \in C} \vartheta[{}^{a/l}_0]\,(0, l\,\Omega) \cdot \mathfrak{z}_J \cdot \det(-l\,\Omega)^{1/2}}{|C| \cdot \vartheta[{}^o_o]\,(0,\Omega) \cdot \mathfrak{z}_J \cdot \det(-\Omega)^{1/2}}$$

$$= \frac{\vartheta_C(\Omega)}{\vartheta[{}^o_o]\,(0,\Omega)} .$$

The last equation holds because $g = 2d$ where $d = \dim_{\mathbb{F}_1} C$ and $|C| = l^d$.

References

A L. Auslander: Lecture notes on nil-theta functions, AMS, CBMS, no. 34, 1977

C Ching-li Chai: Compactification of Siegel moduli schemes, London Math. Soc. Lecture Notes Series, 1o7, Cambridge Univ. Press, 1985

M D. Mumford: Tata lectures on theta I, Birkhäuser, Boston, 1983

O H. Opolka: The finite Fourier transform and theta functions, Proceedings
o of the AAECC-4 conference, Grenoble, 1985; Springer Lecture Notes in computer science, Vol. 229, p. 156 - 166

T R. Tolimieri: The algebra of the finite Fourier-transform and coding theory, AMS Transactions, 287, 1985, 253 - 273

H. Opolka
Math. Institut
Bunsenstr. 3-5

D-3400 Göttingen

ON COMPUTERS AND MODULAR REPRESENTATIONS OF $SL_n(K)$

by

Marilena PITTALUGA and Elisabetta STRICKLAND

Dipartimento di Matematica
II Università di Roma (Tor Vergata)
Via Orazio Raimondo 00173 Roma
Italy

0. Introduction

In this paper we present an algorithm that allows to compute the character and the dimension of the irreducible representations of $SL_n(K)$, K being an infinite field.

The algorithm has been implemented on a computer system, using language MUSIMP (a version of MULISP). The program runs on an IBM PC, and although the performance on such a machine is not excellent, it still allows to work out in a reasonable amount of time examples that cannot be dealt with by hand.

The problem of determining the dimension of the irreducible representations of $SL_n(K)$ is completely solved in characteristic zero; in this case there is an explicit formula (Weyl dimension and character formula). In characteristic p instead, the problem is almost completely open. Clausen [C2] has given an algorithm for the case of the symmetric and general linear group.

It is well known that, if K is an infinite field, the irreducible representations of $SL_n(K)$ are characterized by their highest weights. We briefly recall the terminology:

consider the maximal torus $D_n(k) \subset SL_n(K)$ consisting of diagonal matrices of determinant one; if V is a representation of $SL_n(K)$ and $v \in V$, v is called a *weight vector* if it is a simultaneous eigenvector for all elements of $D_n(K)$. The corresponding eigenvalues will give rise to a character (i.e. a group homomorphism) $\chi: D_n(K) \to K^\times$, called the *weight* of the vector v. In formula this can be summarized by:

$$tv = \chi(t)v \qquad \forall\, t \in D_n(K).$$

The weights of $SL_n(K)$ have a natural total order, so it makes sense to talk of the highest weight appearing in a representation. The basis of this work consists in the following classical result:

Theorem. *If K is an infinite field, each irreducible representation of $SL_n(K)$ has a unique (up to scalars) highest weight vector, and non isomorphic representations have different highest weights.*

Therefore the problem of computing character and dimension of all irreducible representations of $SL_n(K)$ reduces to the problem of determining character and dimension of an irreducible representation of $SL_n(K)$ with given highest weight. We start by recalling the results in characteristic zero.

1. Schur modules

We recall that a *Young tableau* $\lambda = (\lambda_1, \ldots, \lambda_s)$ is a sequence of integers with $\lambda_1 \geq \lambda_2 \geq \cdots \geq \lambda_s \geq 1$. We will be interested only in diagrams with $\lambda_1 < n$. Pictorially we think of a Young diagram as an array consisting of λ_1 boxes in the first row, λ_2 boxes in the second row and so on. For example the diagram $(4,4,3,1,1)$ is represented as:

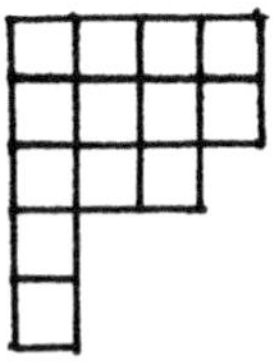

If we fill a Young diagram λ with integers out of $\{1, \ldots, n\}$ we obtain what is called a *Young tableau* of shape λ. A Young tableau can be interpreted as a product of minors of a matrix in the following way: let R be the polynomial ring over $\mathbf{Z}$ in the variables X_{ij} $(i = 1, \ldots, n,\ j = 1, \ldots, n-1)$ and let $X = (X_{ij})$ be the generic matrix over R with entries X_{ij}. Then a row $[i_1, \ldots, i_k]$ of a Young tableau can be interpreted as the determinant of the minor of X consisting of the first k columns and the rows of indices $i_1, \ldots, i_k$ (in this order). The Young tableau itself is then interpreted as the product of the determinants representing each of its rows.

We define the Schur module V_λ as the $\mathbf{Z}$–linear span of all Young tableaux of shape λ. This set of generators for V_λ is not linearly independent, a linearly independent subset being given by the subset of *standard tableaux*, where a standard tableau is a Young tableau in which the rows are increasing sequences and the columns are non–decreasing sequences. For example,

1	2	3	5
1	3	4	5
2	3	5	
2			
4			

is a standard tableau of shape $(4, 4, 3, 1, 1)$. We then have:

Theorem (Hodge). *The standard tableaux of shape λ form a* $\mathbf{Z}$*-basis for* V_λ.

Now observe that the group $G = SL_n(\mathbf{Z})$ acts on the ring R by left multiplication, and that V_λ is G–invariant, hence it is a representation of G. If K is any field, we let $V_\lambda(K) = V_\lambda \otimes_{\mathbf{Z}} K$, and we remark that $V_\lambda(K)$ is actually an $SL_n(K)$–module; moreover every Young tableau in $V_\lambda(K)$ is a weight vector.

2. Irreducible modules in arbitrary characteristic

In V_λ we consider two special standard tableaux: the 'canonical tableau' v_λ, whose i^{th}-row is given by $[1, 2, \ldots, \lambda_i]$, and the 'anti–canonical tableau' u_λ, whose i^{th}-row is given by $[n - \lambda_i + 1, \ldots, n - 1, n]$. If K is an infinite field, v_λ and u_λ are respectively the highest and lowest weight vectors for $V_\lambda(K)$.

The following proposition (a proof of which can be found in [C1]) is crucial for our algorithm:

Proposition. *If K is an infinite field, v_λ and u_λ belong to all non–zero $SL_n(K)$–invariant submodules of $V_\lambda(K)$.*

The proposition implies that the intersection of all non–zero $SL_n(K)$–invariant submodules of $V_\lambda(K)$ is non trivial, therefore $V_\lambda(K)$ is undecomposable. Since $SL_n(K)$ is linearly reductive in characteristic zero, we have:

Corollary. *If K is a field of characteristic zero, $V_\lambda(K)$ is an irreducible $SL_n(K)$–module.*

In the general case, we define L_λ as the G–submodule generated by v_λ and we set $L_\lambda(K) = L_\lambda \otimes_{\mathbf{Z}} K$. Then $L_\lambda(K)$ is the $SL_n(K)$–submodule generated by v_λ and we have:

Corollary. *If K is any infinite field, $L_\lambda(K)$ is an irreducible $SL_n(K)$–module.*

Therefore $L_\lambda(K)$ is an irreducible representation with highest weight vector v_λ, so the $L_\lambda(K)$'s are all the irreducible representations of $SL_n(K)$. We must now find explicitly the dimension of $L_\lambda(K)$.

To do this we will express L_λ as the image of an integral linear map whose matrix we can compute explicitly. The dimension of $L_\lambda(K)$ will then be the rank of this matrix in characteristic p (p being the characteristic of K). To know the dimension of $L_\lambda(K)$ for all characteristics at once, we just have to compute the invariant factors of the matrix we have found.

3. Finding the basic invariant

Let $\lambda = (\lambda_1, \ldots, \lambda_s)$ be a Young diagram, we define the Young diagram *adjoined* to λ as $\widetilde{\lambda} = (\widetilde{\lambda}_1, \ldots, \widetilde{\lambda}_s)$ where $\widetilde{\lambda}_i = n - \lambda_i$. Thus λ and $\widetilde{\lambda}$ fit into a rectangle of width n and height s:

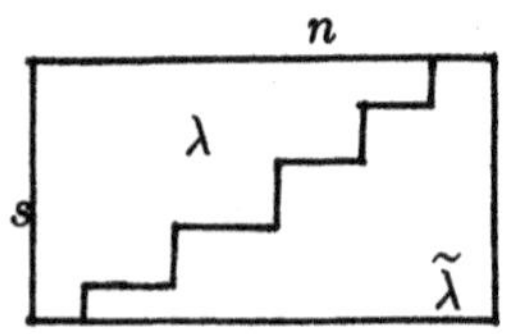

The basic ingredient of the algorithm is the following:

Proposition. $(V_\lambda \otimes V_{\widetilde{\lambda}})^G \neq 0$

In order to prove the proposition (and to exhibit an explicit invariant) we will embed $V_\lambda \otimes V_{\widetilde{\lambda}}$ into a polynomial ring: we can embed $V_{\widetilde{\lambda}}$ in the ring $\widetilde{R} = \mathbf{Z}[X_{ij}]$ ($i = 1, \ldots, m$,

$j = n, \ldots, 2n-2$); in this way we obtain a G–equivariant embedding:

$$V_\lambda \otimes V_{\widetilde{\lambda}} \longrightarrow R \otimes \widetilde{R} = S = \mathbf{Z}[X_{ij}] \qquad \left(\begin{smallmatrix} i=1,\ldots,n \\ j=1,\ldots,2n-2 \end{smallmatrix}\right).$$

There is a natural invariant in S: consider a rectangle of width n and height s, in which we have cut out λ and $\widetilde{\lambda}$ as before:

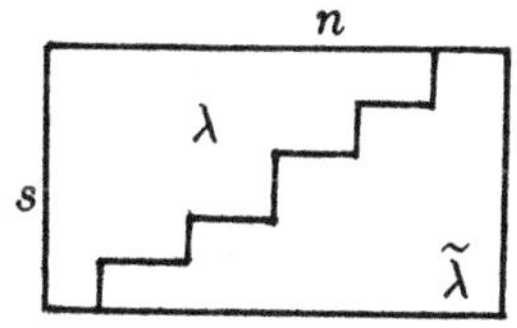

Filling λ canonically with the integers $1, \ldots, n$ and $\widetilde{\lambda}$ canonically with the integers $n, \ldots, 2n-2$, we obtain a Young tableau D of shape $(\overbrace{n, n, \ldots, n}^{k \text{ times}})$ whose i^{th}-row will be $[1, 2, \ldots, \lambda_i, n, n+1, \ldots, n + \widetilde{\lambda}_{n-i}]$. We will think of this row as of the determinant of the maximal order minor of $Y = (X_{ij})$ ($i = 1, \ldots, n$ $j = 1, \ldots, 2n-2$) consisting of columns $1, \ldots, \lambda_i, n, \ldots, n + \widetilde{\lambda}_{n-i}$. Being the product of maximal order minors, the Young tableau D will then represent an invariant in S.

To see that $D \in V_\lambda \otimes V_{\widetilde{\lambda}}$ and to express D explicitly as an element of $V_\lambda \otimes V_{\widetilde{\lambda}}$ we use the Laplace expansion for the determinant. Namely, if A is an $n \times n$ matrix and $1 \leq k \leq n-1$ is an integer, we have:

$$\det(A) = \sum_S \epsilon(S) \alpha_S(1, \ldots, k) \alpha_{\overline{S}}(k+1, \ldots, n)$$

where S runs over all (ordered) k–elements subsets of $\{1, \ldots, n\}$, $\overline{S}$ is the complement of S in $\{1, \ldots, n\}$, $\alpha_S(i_1, \ldots, i_k)$ is the determinant of the minor taken from columns $i_1, \ldots, i_k$ and from the rows indexed by the elements of S, and $\epsilon(S)$ is the sign of the permutation obtained by listing the elements of S and $\overline{S}$ one after the other. (Remark that for $k = 1$ we obtain the usual Laplace expansion over the first column.)

Now we use Laplace expansion with $k = \lambda_i$ for the ith-row of the tableau D; we obtain a sum of products of a minor taken from the first λ_i columns times a minor taken from columns $\{n, \ldots, n + \widetilde{\lambda}_i - 1\}$. Multiplying out these expansions we obtain D as a sum of products of a Young tableau of shape λ times a Young tableau of shape $\widetilde{\lambda}$ (embedded in $\widetilde{R}$). Therefore we obtain an expression:

$$D = \sum_{i,j} \alpha_{ij}\, e_i \otimes f_j,$$

where $\alpha_{ij} \in \mathbf{Z}$, $\{e_i\}$ is the basis of standard tableaux in V_λ and $\{f_j\}$ is the basis of standard tableaux in $V_{\tilde{\lambda}}$.

We remark in passing that the matrix (α_{ij}) is a symmetric matrix.

4. The relation between L_λ and the invariant D

Having found an explicit invariant in $V_\lambda \otimes V_{\tilde{\lambda}}$, we remark now that $V_\lambda \otimes V_{\tilde{\lambda}}$ is isomorphic, as a G-module, to $\operatorname{Hom}(V_{\tilde{\lambda}}^*, V_\lambda)$ and therefore $(V_\lambda \otimes V_{\tilde{\lambda}})^G \simeq \operatorname{Hom}_G(V_{\tilde{\lambda}}^*, V_\lambda)$. It follows that our invariant D will represent a G–equivariant map $\phi_D : V_{\tilde{\lambda}}^* \to V_\lambda$.

We let $W_\lambda = V_{\tilde{\lambda}}^*$ (W_λ is called the *Weyl module*) and we give W_λ the basis dual to the basis of standard tableaux in $V_{\tilde{\lambda}}$. In the chosen bases the map ϕ_D represented by the matrix (α_{ij}) introduced above.

To conclude we simply have to show that the image of ϕ_D is exactly L_λ. To do this it is enough to show that v_λ belongs to the image of ϕ_D. Denote by w_λ the basis element dual to the anti–canonical vector $u_{\tilde{\lambda}} \in V_{\tilde{\lambda}}$, then we want to check that $\phi_D(w_\lambda) = v_\lambda$. But w_λ is a weight vector with the same weight as v_λ, and v_λ is the unique (up to scalars) vector in V_λ with that weight, hence $\phi_D(w_\lambda)$ must be a scalar multiple of v_λ. Now it is easily seen that multiplying out the Laplace expansions of the rows of D we obtain the element $v_\lambda \otimes u_{\tilde{\lambda}}$ exactly once, and neither v_λ nor $u_{\tilde{\lambda}}$ can come out from straightening the other vectors appearing (straightening a tableau does not alter the weight); therefore the scalar multiple can only be v_λ itself.

We conclude with a couple of trivial observations which in practise greatly simplifies machine computations in our problem. If χ is a character of $D_n(K)$ and V is a representation of $SL_n(K)$, we denote by $V_{(\chi)}$ the space of weight vectors of V of weight χ. Then the map ϕ_D, being $SL_n(K)$–equivariant, will map $W_\lambda(K)_{(\chi)}$ to $V_\lambda(K)_{(\chi)}$. Since the bases we chose consist of weight vectors, the matrix of ϕ_D will be in block diagonal form, each block being the matrix of the map $\phi_D: W_\lambda(K)_{(\chi)} \to V_\lambda(K)_{(\chi)}$.

Moreover the symmetric group S_n acts on the weights, so it permutes the spaces $W_\lambda(K)_{(\chi)}$ (resp. $V_\lambda(K)_{(\chi)}$) and this permutation can be seen, up to sign, as a translation by some element of G. Therefore if χ and χ' are two characters which lie in the same S_n–orbit, we have a commutative diagram:

$$\begin{array}{ccc} W_\lambda(K)_{(\chi)} & \xrightarrow{\cong} & W_\lambda(K)_{(\chi')} \\ \Big\downarrow \phi_D & & \Big\downarrow \phi_D \\ V_\lambda(K)_{(\chi)} & \xrightarrow{\cong} & V_\lambda(K)_{(\chi')} \end{array} ;$$

in particular two blocks relative to two weights belonging to the same S_n–orbit will be equal.

5. The algorithm

We will now summarize the algorithm.

Step 1. Generate a set of representatives of the S_n–orbit of weights.

Step 2. For each row of the invariant D, write its Laplace expansion (call these expansions $\mu_1, \ldots, \mu_s$).

Now, for each element χ of the set of representatives we chose in Step 1, perform the following operations:

Step 3. Compute the cardinality m_χ of the S_n–orbit of χ. This will be the multiplicity with which the block relative to χ appears in the matrix of ϕ_D.

Step 4. From the product $\mu_1 \ldots \mu_s$ pick out those elements $v \otimes u$ where v has weight χ, and add these elements up. We obtain an expression of the form:

$$D = \sum_h v_h \otimes \overline{v}_h,$$

where $\overline{v}$ denotes the tableau of shape $\widetilde{\lambda}$ whose i^{th}-row consists of the complement in $\{1, \ldots, n\}$ of the $(n-i)^{\text{th}}$-row of v.

Step 5. Using the straightening algorithm (see for example [DEP1]), express each v_h as a linear combination of standard tableaux:

$$v_h = \sum_j \alpha_{ij} e_j \,,$$

where e_j will necessarily have weight χ. Now observe that the transformation $v \mapsto \overline{v}$ is linear and maps standard tableaux to standard tableaux. Therefore:

$$D = \sum_{j,k} \Big(\sum_h \alpha_{hj} \alpha_{hk} \Big) e_j \otimes \overline{e}_k \,.$$

The matrix $B_\chi = (\beta_{jk})$ where $\beta_{jk} = \sum_h \alpha_{hj}\alpha_{hk}$ is the block of ϕ_D relative to the weight χ.

Step 6. Compute the invariant factors of B_χ and let $s_\chi(p)$ be the number of these which are not divisible by p (p being a prime).

Step 7. For each prime p the dimension of $L_\lambda(K)$ in characteristic p is obtained by multiplying $s_\chi(p)$ times the multiplicity m_χ and adding these up over all representatives χ.

References

[C1] M. Clausen *Letter place algebras and a characteristic–free approach to the representation theory of the general linear group and symmetric groups I and II*, Advances in Math. **33** (1979), 161–191 and **38** (1980), 152–177.

[C2] M. Clausen: *Letter–Place–Algebren und ein characteristik–freier Zugang zur Darstellungstheorie symmetrischer und voller linearer Gruppen*, Bayreuther Mathematische Scriften, Heft **4** (1980), 1–133.

[DEP1] C. De Concini, D. Eisenbud, C. Procesi: *Young diagrams and determinantal varieties*, Inv. Math. **56** (1980), 129–195.

[DEP2] C. De Concini, D. Eisenbud, C. Procesi: *Hodge algebras*, Astérisque (1982).

[DP] C. De Concini, C. Procesi: *A characteristic–free approach to invariant theory*, Advances in Math. **21** (1976), 330–354.

[DRS] P. Doubilet, G. Rota, J. Stein: *On the foundations of combinatorial theory IX*, Studies in Appl. Math. **53** (1974).

[H] W.V.D. Hodge: *Some enumerative results in the theory of forms*, Proc. Camb. Phil. Soc. **39** (1943), 22–30.

[W] H. Weyl: *The classical groups*, Princeton Univ. Press, Princeton N.J. (1946).

CONSTRUCTION OF SDMC CODES
OVER GF(2) OR GF(3)

A. Poli

AAECC/LSI
University P. Sabatier
118 route de Narbonne
31072 TOULOUSE cédex
(France)

RESUME On donne l'algorithme complet permettant de construire tous les codes bordés SDMC, à coefficients dans GF(2) ou dans GF(3). Quelques démonstrations nouvelles sont également développées.

SUMMARY We give a complete algorithm to construct every possible bordered SDMC codes with coefficients in GF(2) or GF(3). Several new proofs are developed too.

INTRODUCTION

The binary and ternary Golay codes, the QR codes all have a generator matrix of a particular type. These matrices are very convenient for construction, decoding, and evaluation of minimum distance.
Such matrices were studied by J. Leech (5), M. Karlin (3) (4), F.J. McWilliams (6), G.M.F. Beenker (1).

More general generator matrices were studied by M. Ventou (14), A. Poli (11), C. Rigoni (13).

Here we recall results that we have already published, only giving the proofs of new ones. We also give here an algorithm to construct every possible bordered SDMC codes over GF(2) or GF(3).

PART I : ALGEBRAIC TOOLS

Definition 1

A multi-circulant, not bordered, code over GF(q) ($q = p^r$, p prime) has, by definition, a kx2k generator matrix $[IQ']$, Q' is a kxk matrix, element of the group algebra $GF(q) \langle\langle M_{r_1} \rangle \otimes \ldots \otimes \langle M_{r_n} \rangle\rangle$.

$\langle M_{r_i} \rangle$ is the cyclic group generated by the $r_i x r_i$ matrix M_{r_i} defined by: $M_{r_i} = \begin{bmatrix} 010\ldots0 \\ 0010\ldots0 \\ \ldots\ldots \\ 0\ldots01 \\ 10\ldots0 \end{bmatrix}$. $\otimes$ means the tensor product.

Example 1

$q=3,\ n=2,\ r_1=r_2=2.$

$$Q' = M_2 \otimes M_2 + 2M_2^2 \otimes M_2^2$$

$$Q' = \begin{bmatrix} 2&0&0&1 \\ 0&2&1&0 \\ 0&1&2&0 \\ 1&0&0&2 \end{bmatrix}.$$

Definition 2

A multi-circulant bordered code over GF(q) has, by definition, a (k+1)x2(k+1) generator matrix of the following type:

$$\begin{bmatrix} a & 0\ldots0 & c & 1\ldots1 \\ b & & d & \\ \vdots & I & \vdots & Q' \\ b & & d & \end{bmatrix}$$

, Q' of the same type as in Definition 1.

Example 2

$q=2,\ n=1,\ r_1=3.$

$$Q' = M_3 + M_3^2.$$

We have:
$$\begin{bmatrix} 1&0&0&0&0&1&1&1 \\ 0&1&0&0&1&0&1&1 \\ 0&0&1&0&1&1&0&1 \\ 0&0&0&1&1&1&1&0 \end{bmatrix}$$

Our goal is to construct multi-circulant codes which are also self dual (Self Dual Multi-Circulant).

SDMC codes are defined by their generator matrix. We will use polynomials.

LEMMA 1 $GF(q)\langle\langle M_{r_1}\rangle \otimes .. \otimes \langle M_{r_n}\rangle\rangle$ is isomorphic to the algebra A:

$A = GF(q)[X_1,..,X_n]/(X_1^{r_1}-1,...,X_n^{r_n}-1)$.

∇ $M_{r_1}^{i_1} \otimes ... \otimes M_{r_n}^{i_n}$ corresponds to $X_1^{i_1}..X_n^{i_n}$ Δ.

We will denote by Q the polynomial which corresponds to the matrix Q'. We will denote by τ the involutive automorphism which corresponds to the transposition of matrix.

Set $k = r_1+..+r_n$, with $r_i = m_i s_i$ (m_i a power of p, s_i relatively prime to p).

Set $m = \max(m_1,..,m_n)$.

PROPOSITION 1

1- Not bordered SDMC codes

1-1 Over GF(2) Every code has exactly one generator matrix [IQ'] such that $1+Q\tau(Q)=0$ holds.

1-2 Over GF(3) There no such code does exist.

2- Bordered SDMC codes

2-1 Over GF(2) The code length $2(k+1)$ is congruent to 0 mod 4. Every code has exactly one generator matrix of the following form:

$$\begin{bmatrix} a0...0 & a+1 & 1...1 \\ 1 & & 1 & \\ \vdots & I & \vdots & Q' \\ 1 & & 1 & \end{bmatrix}$$ with a in GF(2), $1+Q\tau(Q)=0$.

2-2 Over GF(3) The code length $2(k+1)$ is congruent to 0 mod 4. modulo 12. Every code has exactly one generator matrix of the type:

$$\begin{bmatrix} a0...0 & c1...1 \\ x & & y & \\ \vdots & I & \vdots & Q' \\ x & & y & \end{bmatrix}$$ with $a^2+c^2+k=0$; $1-j.1/k+Q\tau(Q)=0$; $x=1/k(\pm c)$; $y=1/k(\mp a)$.

(j is the sum of all (normalized) monomials in A)

∇ Proofs are developped in (11) Δ.

Example 3 Example 1: not a SDMC code; Example 2: yes.

So, to construct SDMC codes, one has to construct polynomials (in the algebra A) verifying $1+Q\tau(Q)=0$, or verifying $1-j.1/k + Q\tau(Q)=0$.

To do that some properties of A are necessary. We give them in the following proposition.

Let μ_i be a root of $X_i^{s_i}-1$ $(1\leq i\leq n)$.

Consider a n-tuple $(\mu_1,...\mu_n)$.

Define $p_1(X_1) = Irr(\mu_1, GF(q))$ as the minimal polynomial of μ_1 over GF(q).

Define $W_i(\mu_1,\ldots,\mu_{i-1},X_i)=\mathrm{Irr}(\mu_i,GF(q)(\mu_1,\ldots,\mu_{i-1})$ $(2\leq i\leq n)$.
Finally define $W_i(X_1,\ldots,X_i)$ (or W_i) as obtained from $W_i(\mu_1,\ldots,X_i)$ by substituting X_j to μ_j $(1\leq j\leq i-1\leq n-1)$.

One useful tool for our constructions is the decomposition of A as a direct sum of sub-algebras.
Each sub-algebra is related to one n-tuple $(\mu_1,\ldots,\mu_n)$. Suppose that this n-tuple defines the n-tuple $(p_1,W_2,\ldots,W_n)$, then we have the following properties.

PROPOSITION 2

1- The set of all primitive idempotents in A is in a one-to-one correspondence with all n-tuples $(p_1,W_2,\ldots,W_n$. By convention, e_N will denote the primitive idempotent which is associated with $(X_1-1,\ldots,X_n-1)$. e_N is equal to j.1/k (for bordered SDMC codes over GF(2), or over GF(3) k is not zero).

2- A is equal to the direct sum $(e_1)\oplus\ldots\oplus(e_N)=A_1\oplus\ldots\oplus A_N$.
Moreover A is ring isomorphic with the cartesian product $B_1\mathrm{x}\ldots\mathrm{x}B_N$.
Each B_i is of the type $GF(q')[Z_1,\ldots,Z_N]/(Z_1^{m_1},\ldots,Z_n^{m_n})$. B_i is ring isomorphic to (e_i). If e_i is associated to $(p_1,W_2,\ldots,W_n)$ then the corresponding field $GF(q')$ is the field $GF(q)[X_1,\ldots,X_n]/(p_1,\ldots,W_n)$.

3- Let τ' be the automorphism in the cartesian product which corresponds to τ in A.
If $\tau'(B_i)=B_i$ then $q'=q^{2\delta_i}$, and $\tau'(x)=x^{q^{\delta_i}}$ $(=\bar{x})$ for every x in $GF(q')$,
or $q'=q$ and $\tau'(x)=x$ for every x in $GF(q')$.

4- There exist a ring isomorphim from B_i to (e_i) defined by :
$X_j \longrightarrow X_j^m e_i$; $Z_k \longrightarrow p_k e_i$.

∇ Proofs are given in (10) Δ

The polynomial Q we want to construct can be expressed as $Q=Q_1+\ldots+Q_N$ in A. To each Q_i in A_i there corresponds one q_i in B_i. We will determine the q_i's.
Two cases are possible depending on whether or not B_i is invariant under τ'.

PROPOSITION 3

1- If $\tau'(B_i)=B_j\neq B_i$ then $(q_i,q_j)=(u,-\tau'(u^{-1}))$, for every invertible element u in B_i.

2- If $\tau'(B_i)=B_i$ then q_i has to be a solution of $q_i\tau'(q_i)=-1$.

∇ See (12), (13) Δ

The next Proposition gives us a practical manner to know what case holds.

PROPOSITION 4

The algebra B_i, associated to $(p_1,W_2,..,W_n)$, is left invariant under τ' iff the two following conditions are satisfied :

1- $p_1,p_2,..,p_n$ are self reciprocical (with respective degrees $d_1,d_2,...,d_n$).

2- Each d_i is 1 or $2^l d_i'$, d_i' odd, l does not depend on i.

∇ This proof is new.

As B_i is associated to some $(p_1,W_2,..,W_n)$ it is also associated to the corresponding n-tuple $(\mu_1,\mu_2,...,\mu_n)$. $\tau'(B_i)$ is associated to $(\mu_1^{-1},...,\mu_n^{-1})$ (11).

Suppose that $\tau'(B_i)=B_i$.

There exist λ such that $\mu_1^{-1}=\mu_1^{q^\lambda},....,\mu_n^{-1}=\mu_n^{q^\lambda}$.

Consequently the p_i's are self reciprocical.
Then d_i is 1 or is even. When d_i is even $\mu_i^{-1}=\mu_i^{q^{d_i/2}}$ holds. Then, for each even d_i one has: $2\lambda= d_i(1+2\omega_i)$. For d_i and d_j both even one has: $d_i(1+2\omega_i)= d_j(1+2\omega_j)$. Observe that $1+2\omega_i$ and $1+2\omega_j$ are odd, and conclude that 2- holds

Suppose that 1- and 2- hold.

Set $2^l t$ for the LCM of all the even d_i's.

Set $\lambda= 2^{l-1}t$. Prove that $\mu_i^{-1}= \mu_i^{q^\lambda}$ $(1\leq i\leq n)$ Δ.

The calculations we develop in our proposed algorithm (Part 2) deal with particular bases in B_i (when B_i is invariant).

PROPOSITION 5

1- $q'\neq q$ There exist a basis of the GF(q')-vector space B_i each of whose elements is left invariant by τ'.

2- $q'=q$, q odd There no invariant basis does exist. But there exist a semi-invariant basis, i.e. each non invariant element is sent to its opposite.

3- $q'=q$, q even There no invariant (nor semi-invariant) basis does exist.

∇ The proofs for existence are developped in (11). Here, we only develop the (new) proof of non-existence.

1- Let a_i be the constant term in $\tau'(Z_i)$. There exist b_i in GF(q')* such that $b_iZ_i+\bar{b}_i\tau'(Z_i)$ (=f_i) has the coefficient of Z_i not zero. The desired basis is $\{f_1^{i_1}..f_n^{i_n},\ 0\leq i_j<m_j,\ 1\leq j\leq n\}$.

2- Set $f_i = Z_i - \tau'(Z_i)$. The desired basis is $\{f_1^{i_1}..f_n^{i_n}\}$.

Now we prove that no invariant basis does exist.

Suppose $s(Z_1,..,Z_n)$ be invariant. Let b be the coefficient of Z_1.
One has: $s=bZ_1+ s'$, $\tau'(s)=s=b\tau'(Z_1)+\tau'(s')$.
As B_i is invariant (for q'=q) then p_1 is X_1-1 or X_1+1.

2-1 $p_1=X_1-1$ Recall that $r_i=m_is_i$ (see after Lemma 1).
$\tau'(Z_1)$ corresponds to $(X_1^{r_1-1}-1)e_i$ in A (Proposition 2,4-).
Prove that: $X_1^{r_1-1}-1\equiv X_1^{m_1-1}-1$ modulo $X_1^{m_1}-1$. At last:
$X_1^{m_1-1}-1= (X_1-1)((X_1-1)^{m_1-1}-1)/((X_1-1)+1))$.

One deduce, in B_i: $\tau'(Z_1)=-Z_1+Z_1^2-...+(-1)^{m_1-1}Z_1^{m_1-1}$.
Then: $\tau'(s)=s= -bZ_1 + s''$. Each term in s'' has degree in Z_1 strickly greater than 1. Then b=-b, and b=0 (q is odd).

2-2 $p_1=X_1+1$ In the same way: $\tau'(Z_1)=-Z_1-Z_1^2-..-Z_1^{m_1-1}$,and b=0.

The vector space generated by the set of all invariant polynomials s does not contain Z_1. Then no invariant basis does exist.

3- As q is even there is no difference between invariant or semi-invariant basis.
Suppose $s'(Z_1,..,Z_n)$ be invariant.
Let $s=bZ_1+cZ_1^2+s'$ (s' without Z_1 nor Z_1^2).
One has: $\tau'(Z_1)=Z_1+Z_1^2+..+Z_1^{m_1-1}$. Then

$$\tau'(s)=s=b(Z_1+Z_1^2+..)+c(Z_1^2+Z_1^4+..)+\tau'(s'),\ c=b+c,\ b=0.$$

Then no invariant basis does exist. Δ

PART II : THE ALGORITHM TO CONSTRUCT SDMC CODES

For sake of simplicity we only develop the algorithm for bordered SDMC codes over GF(2) or GF(3). For not bordered binary codes correspond to the much more complicated case "q'=q, q even".

Let us give the different steps of the algorithm which is available for every characteristic p.

1- Construction of $m_1,m_2,..,m_n$, $X_1^{s_1}-1,..,X_n^{s_n}-1$.

2- Factorize $X_i^{s_i}-1$. Set p_i for an irreducible factor of $X_i^{s_i}-1$.

3- If all possible n-tuples $(p_1,p_2,..,p_n)$ have been considered then END, else select one of the remainding ones.

4- Factorize p_2 over $GF(q)(\mu_1)$. Obtain $W_2(X_1,X_2)$. Then factorize p_3 over $GF(q)(\mu_1,\mu_2)$. Obtain $W_3(X_1,X_2,X_3)$....... We obtain one n-tuple $(p_1,W_2,..,W_n)$ which is associated to some A_i.

5- Construct e_i, using Poli's algorithm (10).

6- In this step we deduce every n-tuple $(p_1,W'_2,..,W'_n)$ associated to every A_j, and the corresponding e_j.

$W'_2(\mu_1,X_2)$ is $Irr(\mu'_2,GF(q)(\mu_1))$, μ_2 a root of p_2,.....

.....$W'_n(\mu_1,\mu'_2,..,\mu'_{n-1},X_n)$ is $Irr(\mu'_n,GF(q)(\mu_1,\mu'_2,..,\mu'_{n-1}))$ μ'_n root of p_n

Using the lexicographical ordering on the set of all n-tuples $(l_1,..,l_n)$ s.t. $0 \leq l_i < GCD(LCM(d_1,..,d_{i-1}),d_i)$, $(2 \leq i \leq n)$ one proceeds as follows :

6-1 (Polynomials)

For i from 2 to n: W'_i is $W_i(X_1^{q^{-l_i}},X_2^{q^{l_2-l_i}},..,X_{i-1}^{q^{l_{i-1}-l_i}},X_i)$ reduced modulo $(p_1,W'_2,..,W'_{i-1})$.

6-2 (Idempotents)

Calculate $e_i(X_1,X_2^{q^{l_2}},...,X_n^{q^{l_n}})$.

7- (Construction of the solutions in B_i)

7-1 (B_i not invariant)

Construct u, invertible in B_i. u is any polynomial of the type $u=\beta+v(Z_1,..,Z_n)$, β in $GF(q')^*$, v a polynomial with a constant term equal to 0.

Deduce u^{-1}, and then $-\tau'(u^{-1})$ in $\tau'(B_i)$.

7-2 (B_i invariant)

7-2-1 (q'=q, q odd)

That case arises only when $p_i=X_i-1$ $(1 \leq i \leq n)$. For, k is odd (Prop. 1,2-2) and then -1 is not a root.

Then A_i is A_N. Now, each polynomial Q verifying $1-e_N+Q\tau'(Q)=0$ has a zero component in A_N(for, $1-e_N$ has). Then q_N is equal to 0.

7-2-2 (q'=q, q even)

$p_i=X_i+1$ $(1 \leq i \leq n)$, $A_i=A_N$.

k is odd (Prop. 1 ,2-1). Then $m_1=..=m_n=1$, B_N is equal to GF(2).

As Q must verify $1+Q\tau'(Q)=0$ then q_N is equal to 1.

7-2-3 ($q' \neq q$)

7-2-3-1 (Construct the invariant basis (Prop.5,1-)).

7-2-3-1-1 Determine a_i, the coefficient of Z_i in $\tau'(Z_i)$.

7-2-3-1-2 Set $b_i=1$ if $a_i \neq -1$ .If $a_i=-1$, choose b_i as every element in $GF(q') \backslash GF(q)$.

7-2-3-1-3 Construct $f_i=b_iZ_i+\bar{b}_i\tau'(Z_i)$.

7-2-3-1-4 Construct the whole basis.

7-2-3-2 (Solve $g\tau'(g)=-1$ in B_i)

Set (i) for $(i_1,\ldots,i_n)$, $0\le i_j<m_j$, $1\le j\le n$.

One has $(\Sigma g_{(i)}f^{(i)})(\Sigma\bar{g}_{(i)}f^{(i)})=-1$ $(=\Sigma E(i)f^{(i)})$.

It is equivalent to solve: $-1=E(0)=\bar{g}_{(0)}g_{(0)}=g_{(0)}^{q^{\delta_i}+1}$ (See Prop.2 for δ_i).

and : $0=E(s)=g_{(0)}\bar{g}_{(s)}+\bar{g}_{(0)}g_{(s)}-u$, $(s)\neq(0)$.

Remark that u is with only $g_{(i)}$'s such that (i) is less than (s) for the product ordering (we denote (i)P(s)),and that $(i)\neq(0)$, $(i)\neq(s)$. Equations $E(s)=0$ can be solved one after the other using the lexicographic ordering L (For,(i)P(j) implies (i)L(j)).

7-2-3-2-1 (To solve $E(0)=-1$)

One has to determine a primitive root of the equation $X^{q^{\delta_i}+1}+1=0$.(Recall that $2\delta_i=d'$).

To do that one has to construct a primitive element of $GF(q^{2\delta_i})$.

This field is equal to $GF(q)[X_1,\ldots,X_n]/(p_1,W_2,\ldots,W_n)$.

It is needed to know a primitive polynomial P(Y) over GF(q),with degree $2\delta_i$. Then, construct an irreducible factor of P(Y) over GF(q') in the same way as to factorize p_2 over $GF(q)(\mu_1)$ (see 1-3).

Let β be the primitive element.

For q=2 (resp. q=3) $\beta^{2^{\delta_i}-1}$ (resp. $\beta^{(3^{\delta_i}-1)/2}$) is the desired element.

7-2-3-2-2 (To solve $E(s)=0$, $(s)\neq(0)$).

Set $X=\bar{g}_{(0)}g_{(s)}$. One has to solve $X+\bar{X}=u$.

7-2-3-2-2-1 Construct the matrix M of the linear mapping $X\dashrightarrow X+\bar{X}$, using the basis of all monomials $X_1^{i_1}\ldots X_n^{i_n}$.

7-2-3-2-2-2 Using Gauss' method put M into its reduced echelon form. Then one has directly all possible solutions.

7-2-3-2-2-3 Deduce all $g_{(s)}$', using $g_{(s)}=X(\bar{g}_{(0)})^{-1}$ for all possible $g_{(0)}$'s.

8- (Coming back to A).

Having all possible solutions in the B_i's,one uses the isomorphism described in Prop.2,4-.

9- (Construction of the polynomials Q).

Simply add the polynomials obtained in 8-

CONCLUSION

At the AAECC Lab we have a software which constructs SDMC codes. This software was written by C.Ajonc, in FORTRAN 77. It runs on a Burroughs 7600, a CDC 750, and a DPS 8/70. Note that our proofs are available for more general algebras than the group algebra A.

References

(1) Beenker GMF "On double circulant codes"
TH. Report 80,WSK-04,1980,Univ.Eindhoven.The Need.

(2) Berlekamp ER "Algebraic coding theory"
Mac Graw Hills,NY,1968

(3) Karlin M "New binary coding results by circulants"
IEEE Trans. on Inf.Th.,vol 15,pp797-802,1969.

(4) Karlin M "Decoding of circulant codes"
IEEE Trans. on Inf.Th.,pp797-802,Nov.1970.

(5) Leech J "Some sphere packing in higher space".
Can. J. of Math.,pp657-682,1964.

(6) MacWilliams FJ "Orthogonal circulant matrices over finite fields and how to find them"
J. of Comb. Th.,vol.10,pp1-17,1971.

(7) MacWilliams FJ
Sloane NJA "The theory of error correcting codes"
North Holland Math. Lib.,3nd Edit.,1981.

(8) Malliavin MP "Les groupes finis et leurs représentations complexes"
Masson,NY,1981.

(9) Poli A "Idéaux principaux nilpotents de dimension maximale dans F_qG algèbre de groupe abélien fini".
Comm. in Algebra,12(4),pp391-401,1984.

(10) Poli A "Construction of primitive idempotents for n variable codes"
Lect. Notes, in Comp. Sc., Springer n°228, pp 25-35 1986

(11) Poli A "Multi-circulant codes over F_q"
Submitted to Comm. in Algebra.

(12) Poli A,Rigoni C "Enumeration of self dual 2k circulant codes"
Lect. Notes, in Comp. Sc., Springer n°228, pp61-71, 1986.

(13) Poli A,Rigoni C "Self dual codes 2n circulant over F_q $(q=2^r)$".
Lect.Notes in Comp. Sc., Springer n° 229,pp194-201

(14) Ventou M,Rigoni "Self dual doubly circulant codes"
Discrete Math.,vol.56,pp291-298,1985.

°0°0°0°0°0°

FAST16 : A SOFTWARE PROGRAM FOR FACTORISING POLYNOMIALS OVER LARGE GF(P)

A. Poli, M.C. Gennero
AAECC/LSI Lab., Université P. Sabatier
Toulouse - France

SUMMARY

We propose experimental results, obtained from an algorithm for factorising polynomial over large prime fields GF(p).
Experimental running times are compared with those needed using Macsyma (for the same polynomials, on the same computer).
The proposed algorithm includes original contributions : a strategy based on Stickelberger's theorem, a particular use of the Norm function. and a systematic use of sequences of squares/non-squares in GF(p).

INTRODUCTION

Algorithms for polynomial factorisation can be viewed through a simple scheme.
We first present this scheme as a game (part I), which we then use to examine the classical factorisation methods.
In part II, we give our algorithm, and we discuss some particular points.
Finally, in part III we give the experimental results we have obtained : running times, maximal length sequences of squares/non-squares in GF(p) for p large (up to 999983).

In our conclusion we propose a conjecture on maximal length sequences of squares/non-squares, and following Tarjan and Hopcroft we consider the data management problem for factorisation.

PART I : A SIMPLE SCHEME TO REPRESENT FACTORISATION

1.1. A GAME

Let us consider various levels of a game.

LEVEL 1 : Let A be the finite field GF(16), the extension field of degree 4 over GF(2).

One element x is chosen in A. You don't know what element x is. From x you have to construct the 0 element, by performing additions and multiplications in A.
You also have the possibility to randomly choose another element z in A.
Every test, to determine whether the element you are considering is zero or not, has cost of t. Addition has no cost. Let m be the cost of one multiplication.

The object of the game is to obtain zero, with minimal cost.

HOW TO WIN ?

Several strategies are possible, as described in the following.

Strategy 1 : $x \to x^2 \to x^4 \to x^8 \to x^{12} \to x^{14} \to x^{15} \to x^{15}-1 \to$ Test

- Probability of winning : $p=15/16$ (only the zero element is not a root of $x^{15}-1$)
- Mean number of tries : $\bar{n}=16/15$ (Geometric-like distribution)
- Cost for one try : $c=6m+t$ (6 multiplications and one test)
- Mean cost : $\bar{c}=\frac{16}{15}(6m+t)$

Strategy 2 : Choose an element z randomly and do :
$x \to x+z \to (x+z)^5 \to (x+z)^5-1 \to$ Test
One has : $P=1/5$; $\bar{n}=3$; $c=2m+t$; $\bar{c}=3(3m+t)$

Another possibility is :
$x \to x+z \to (x+z)^3 \to (x+z)^3-1 \to$ Test
one has : $P=1/2$; $\bar{n}=5$; $c=2m+t$; $\bar{c}=5(2m+t)$

Strategy 3 : $x \to x^2 \to x^4 \to x^8 \to x+x^2+x^4+x^8 \to$ Test
One has : $P=1/2$; $\bar{n}=2$; $c=3m+t$; $\bar{c}=2(3m+t)$

<u>LEVEL 2</u> : Let A be the cartesian product GF(16) x GF(16).

One element x $(x=(x_1,x_2))$ is randomly chosen in A. You don't know what the element x is.
You have to construct $(0,y_2)$ or $(y_1,0)$ with y_1, y_2 not zero.

As for level 1 you can perform additions and/or multiplications in A. Also you can choose an element z at random.

m and t have the same meaning as previously, but do not necessarily have the same values.

Strategy 1 : $x \to x^2 \to x^4 \to x^8 \to x^{12} \to x^{14} \to x^{15} \to x^{15}-1 \to$ Test
$P=30/16^2$; $\bar{n}=8.53$; $c=6m+t$; $\bar{c}=8.53(6m+t)$

Strategy 2 : Choose z at random, and $x \to x+z \to (x+z)^5 \to (x+z)^5-1 \to$ Test
$P=4/9$; $\bar{n}=9/4$; $c=3m+t$; $\bar{c}=\frac{9}{4}(3m+t)$

Strategy 3 : $x \to x^2 \to x^4 \to x^8 \to x+x^2+x^4+x^8 \to$ Test
$P=1/2$; $\bar{n}=2$; $c=3m+t$; $\bar{c}=2(3m+t)$

The reader will have observed that the first strategy corresponds to the Norm function, and strategy 3 to the Trace function. Strategy 2 is based on the set of multiplicative subgroups of A.

Norm and Trace functions are useful tools for factorising. In the case of an extension field of GF(2), Norm and Trace give 1 or 0 as result. If the result is not zero it is necessarily equal to 1, and so we win in both cases.

For an extension field of GF(2) $(p \neq 2)$, we don't win so directly.

At level 3, we test if we have $(y_1,0)$ or $(0,y_2)$ (y_1, y_2 not zero).

<u>LEVEL</u> 3 : Consider $A=GF(p^r) \times GF(p^s)$ (p prime, r and s are integers).

You may not know the values of r and s. Then, Norm and Trace functions are no longer available. The set of subgroups in A is not known.

A possible strategy is :

a) $x \to y$
b) $x \to x^p \to x^p-y$
c) Test. <u>If</u> the test fails <u>then</u> substitute x^p for x and go back to b)
<u>else</u> : you win

This strategy is known as the "distinct degree factorisation algorithm" (Golomb et al. [8])

Now let us give the basic scheme for factorisation.

1.2 A SIMPLE SCHEME TO REPRESENT FACTORISATION

Now we introduce polynomials

Consider $f(x)$ a square-free polynomial over $GF(p)$.
Let $f(x)=p_1(x) \ldots p_k(x)$. Let d_i be the degree of $p_i(x)$ $(1 \le i \le k)$.
Let $A=GF(p)[x]/(f(x))$.

PROPERTY 1- ° *A is equal to the direct sum :*
$A = (E_1(x)) \oplus \ldots \oplus (E_k(x))$,
where $(E_i(x))$ *is the principal ideal generated by the primitive idempotent* $E_i(x)$ $(1 \le i \le k)$

° *Moreover A is isomorphic to the cartesian product of fields :* $A \cong GF(p^{d_1}) \times \ldots \times GF(p^{d_k})$
Each $GF(p^{d_i})$ *is isomorphic to the corresponding* $(E_i(x))$ $(1 \le i \le k)$.

∇Δ Classical. Recall that $E_i(x)$ is a multiple of $f(x)/p_i(x)$ Δ∇

Let $a(x)$ be an element of $A \setminus \{0\}$. One has $a(x)=a_1(x)E_1(x)+\ldots+a_k(x)E_k(x)$ (with degree of $a_i(x)$ < degree of $p_i(x)$ $(1 \le i \le k)$).

PROPERTY 2- *GCD(f(x),a(x)) is nontrivial iff at least one* $a_i(x)$ *is zero.*

∇Δ Suppose $a_t(x)=0$, for some subscript t.
Then it follows from property 1 that $p_t(x)$ is a factor of $a(x)$.
Now suppose $a_t(x) \neq 0$ $(1 \le i \le k)$.
Then $a(x)$ is a unit in A and, consequently, $GCD(f(x),a(x))$ is trivial Δ∇

The factorisation of $f(x)$ may be viewed through the game described in 1.1, as indicated in the following corollary.
Suppose $a(x)$ is in $A \setminus \{0\}$. Suppose that $(x_1,\ldots,x_k)$ is the corresponding element in $GF(p^{i_1}) \times \ldots \times GF(p^{i_k})$.

COROLLARY- *One has nontrivial factorisation of f(x) iff at least one* x_i *is zero.*

∇Δ Directly from properties 1 and 2 Δ∇

Now we present the scheme for factorising $f(x)$.

A polynomial a(x) is given in A (randomly or not). It corresponds to $(x_1,\ldots,x_k)$ in the cartesian product.

The basic task is to deduce an element $y=(y_1,\ldots,y_k)$ with at least one y_i equal to zero. To do that, it is possible to directly construct such an element. It is also possible to construct first an element $(t_1,\ldots,t_k)$ with t_i in GF(p), and then try to construct a "good" element. Here, testing corresponds to calculate GCD(a(x),y). The test fails iff the result is a constant term.

Classical algorithms for polynomial factorisation may be viewed through this scheme.

1.3 CLASSICAL IDEAS FOR FACTORISING POLYNOMIALS

The main ideas for factorising, which can be found in the literature, are now given.

Golomb et al. [8] (1959).
They calculate $x^{p^e}-x$ in A, and then test. This is known as "distinct degree factorisation" technique. For each value of e the result is the product of all factors of f(x) whose degrees divide e.

Berlekamp [1] (1968).
He solves the equation $z^p-z=0$ in A (Berlekamp's subspace). Every solution is an element $(x_1,\ldots,x_k)$, each x_i in GF(p). If p is not too large Berlekamp proposes to calculate $(x_1,\ldots,x_k)-(1,\ldots,1)$ and to test. If the test fails (GCD equal to a constant) he then substitutes $(x_1,\ldots,x_k)-(1,\ldots,1)$ for $(x_1,\ldots,x_k)$ and iterates.
This algorithm is efficient for small values of p (less than 29 [10, page 423]).

Zassenhaus [17] (1969).
He uses the set of multiplicative subgroups H_j of order $(p^k-1)/j$. He calculates $x=(x_1,\ldots,x_k)-(s,\ldots,s)$, for s randomly chosen in GF(p). Then he calculates x^j-1, and tests.

In the same paper, he proposes to transform the problem of factorising f(x) into the problem of finding the roots of another polynomial g(x). His idea is equivalent to considering the monomial x in A, and the corresponding $(x_1,\ldots,x_k)$ in the cartesian product of fields. The scalars $x_1,\ldots,x_k$ are the desired roots of g(x).

Berlekamp [2] (1970)
In this (nice) paper, he proposes to search the products $h_e(x)$ of factors with same degree e. Then he suggests to transform the problem to factorise $h_e(x)$ into the problem to search the roots of some other polynomial (Zassenhaus's idea). For that he recommends to use Trace function, and some extraneous polynomial to be sure that one obtains a non-trivial factorisation.
He suggests that Stickelberger's theorem should be interesting, as well as sequences of squares/non-squares in large GF(p). He wrote that nothing is known about sequences of squares/non-squares [2 page 732].

Moenk [12] (1977)
He supposes that $p-1=L2^e$ holds, and uses the set of multiplicative subgroups of order 2^i $(0 \leq i \leq e)$ of GF(p).

Rabin [15] (1979)
He uses the same idea as Zassenhaus, but with s is some extension of GF(p).

Cantor-Zassenhaus [6] (1981)
They choose an element $(x_1,...,x_k)$, randomly. They use the Norm function, substract (1,...,1) from $(x_1,...,x_k)$, and test.

Camion [4][5] (1982-83)
He uses elements in Berlekamp's subspace. Using these elements he splits up an idempotent of A into a sum of two orthogonal idempotents. He iterates as long as he obtains primitive idempotents, and then the irreducible factors of f(x).
He proposes to use the Trace function, developping an idea wich resembles extraneous polynomials of Berlekamp [2].

Poli [13] (1983)
He factorises binary polynomials using Berlekamp's subspace. Then he uses a projection from A onto some subalgebra B, to factorise f(x) as soon as possible (even if the factor is not irreducible).
This projection sends the Berlekamp's subspace of A onto that of B.
The corresponding software FAST05 factorises polynomials with degree up to 2100.

PART II : OUR ALGORITHM

Recall that $f(x)$ is square free, and is equal to the product of k irreducible factors $f_1(x),\ldots,f_k(x)$ whose degrees are $d_1,\ldots,d_k$ respectively.

In the following we give :

1- The main steps of our algorithm, with remarks and comments

2- The flowchart

3- Details on the flowchart.

2.1. MAIN STEPS OF OUR ALGORITHM

1- We search one factor of $f(x)$, which is the product of all irreducible polynomials $f_i(x)$ whose degrees is divisor of some integer r. Let $f_{(r)}$ be such a factor.
This step corresponds to the test : "is $x^{p^j}-x$ suitable ?" in the flowchart.

2- Using Stickelberger's result we examine what are the possible factorisations of $f_{(r)}$. Then we try to factorise.

3- If the factorisation provides a non-irreducible factor $g(x)$, then

3.1- We use $\mathrm{Norm}(x^{(p-1)/2})$ and $\mathrm{Norm}(1+x)^{(p-1)/2}$ to try to factorise $g(x)$.

3.2- We use $T(x^i)$ to obtain an element y in $GF(p)[x]/(g(x))$. y corresponds to a k-uple $(y_1,\ldots,y_k)$ in the cartesian product of finite fields (with $y_1,\ldots,y_k$ in $GF(p)$).

3.3- We compute $y^{(p-1)/2}$ and try to factorise. If it fails then we substitute y-1 to y and we iterate 3.3

Points 2- and 3- correspond to "PEELER" in the flowchart.

Remarks and comments : point 1- of the algorithm

Set $d=\min(d_1,\ldots,d_k)$. Using Golomb's idea we calculate $GCD(f(x),x^{p^s}-x)$ for several values of s.

PROPERTY 3- *$f_{(d)}$ will be obtained after d GCDs if we use the sequence $s=1,2,\ldots,$ and on average after $(d+1)/2$ GCDs if we use a sequence $s=r,r+1,\ldots$ where r is randomly chosen.*

∇Δ Direct Δ∇

We also decrease the number of GCDs required, by using Stickelberger's result [1 page 26].

PROPERTY 4- *Using Stickelberger's theorem we need at most :*
$\lfloor n/2 \rfloor - \lceil n/4 \rceil$ GCDs if the number of factors is even
$\lceil n/3 \rceil - \lceil n/6 \rceil$ GCDs otherwise
for factorising polynomials of degree n.

∇Δ Direct Δ∇

Remarks and comments : point 3.1. of the algorithm

In general, it is more expensive to calculate Norm(u) than Tr(u). But, as $x^{(p-1)/2}$ is already calculated (because the Q matrix is obtained), it is interesting to calculate Norm($x^{(p-1)/2}$).

Remarks and comments : point 3.2. of the algorithm

From Berlekamp [2] we know for sure that a factorisation of $f_r(x)$ will be obtained.
Set $t_i = Tr(x^i)$, and suppose that t_i corresponds to a k-uple $(y_1,\ldots,y_k)$. Following Berlekamp [2] we can factorise g(x) by calculating GCD(g(x), t_i), GCD(g(x), t_i-1),... This method is available only for small values of p. We have :

PROPOERTY 5- *The mean number (MN) of GCDs to be calculated for large values of p is equivalent to $p/(s+1)$ (s is the number of irreducible factors of g(x)).*

∇Δ Let i be an integer in the range $\{0,\ldots,p-1\}$. The probability P(i) that every y_j is greater than or equal to i, and at least one y_j is equal to i, is given by :

$$P(i) = ((p-i)/p)^s - ((p-i-1)/p)^s$$

Then $MN = \sum_{i=0}^{p-1} i\,P(i) = p^{-s} \sum_{i=0}^{p-1} i^s = p^{-s} S_s(p-1)$

Observing that : 1/ $n^{-s} S_j(n-1) \to 0$ as $j \le s-2$, and $n \to \infty$
2/ $n^{-s} S_j(n-1) \to s^{-1}$, as $n \to \infty$,

we deduce that $MN\,(s+1)/p \to 1$, as $p \to \infty$ Δ∇

The value of $p/(s+1)$ is an upper bound for MN. This bound is good as confirmed by simulations (MN_e is the experimental value of MN) :
p=14983, s=3 (100000 tests) : MN_e= 3741.7; p/(s+1)= 3745.7
p=100999, s=3 (200000 tests): MN_e= 25182.3; p/(s+1)= 25249.8

p= 999983, s=4 (100000 tests) : MN_e=199760.0; p/(s+1)= 199996.7

For large p it is not possible to use this method. This is the reason why we use exponentiation by (p-1)/2. The problem of the efficiency of this strategy (in mean, or in the worst case) is a very hard one (see Berlekamp [2] page 732).

This problem is equivalent to the following one :

a- Choose at random two elements x_1 and x_2 in GF(p)

b- Test if (x_1-s) and (x_2-s) are both squares or both non-squares (for some integer s). If no you win, if yes then change the value of s and repeat b-

The question is : how many tests do you have to do (on the average, in the worst case) ?

One has some partial answers :

- Camion [5] has proved that there exists in GF(p) a subset S such that $(x_1-s)(x_2-s)$ is a non-square for at least one element s in S. He also proved that #S (the cardinal of S) is less than $[2\log_2 p]$. But he did not give any constructive proof.

- Burgess [3] has proved that the maximal number of consecutive squares in GF(p) for large p is $O(p^{1/4+\varepsilon})$ for every $\varepsilon \geq 0$.

- See also H. Davenport and P. Erdos [7].

Thus, in the worst case the number of trials we have to carry out is greater than or equal to the number (ncs) of consecutive squares in GF(p). But this number may be greater : the worst case may corresponds to two sequences of squares/non-squares $(z_1,\ldots,z_t)$ and $(z'_1,\ldots,z'_t)$ ($z_i z'_i$ is a square) with t greater than ncs.

For s, the sequence of values we have chosen is : s=0,1,2,...

Set m the maximum length of t such that there exist sequences $(z_1,\ldots,z_t)$, $(z'_1,\ldots,z'_t)$ with $z_i z'_i$ a square.

In part III we give a comparison between #S and m for several values of p. It seems interesting to remark that m is not very different from #S.

2.2. THE FLOWCHART

We present now the flowchart of the software program we have formulated from our algorithm. This figure gives only the main steps, and we discuss the processes in part 2.3.

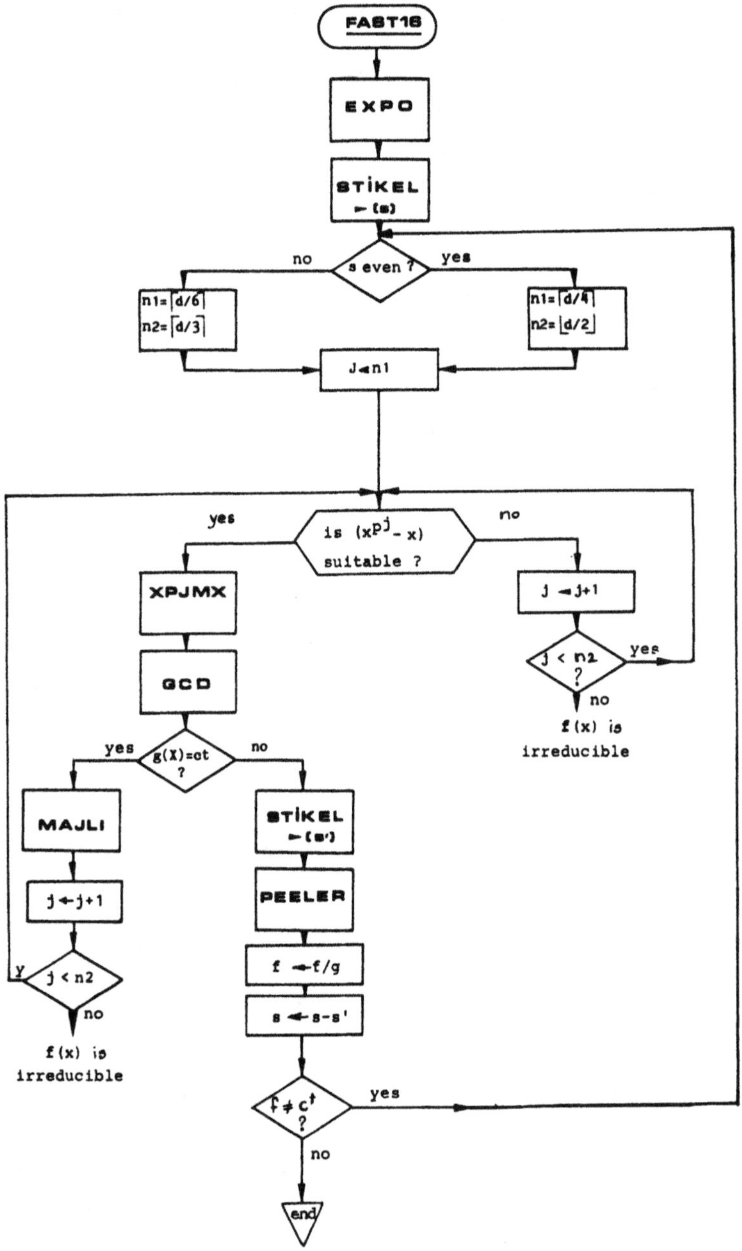

Figure 1: Flowchart of our algorithm for factorising polynomials over GF(p), p prime.

2.3. DETAILS ON THE FLOWCHART

1- EXPO: Construction of the Q matrix (exponentiation by p in A)

2- STIKEL: Determination of the parity **s** of the number of factors.

3- Set : $n_1 = \lceil n/4 \rceil$ and $n_2 = \lfloor n/2 \rfloor$ if s is even
$n_1 = \lceil n/6 \rceil$ and $n_2 = \lceil n/3 \rceil$ if s is odd

4- We search one factor of f(x), which is the product of irreducible polynomials, whose degrees divide the same integer j. Let us call this polynomial g(x).

Set $j=n_1$, do :

4.1- XPJMX : Procedure to compute the polynomial $x^{p^j}-x$, modulo f(x).

4.2- GCD : procedure to compute $g(x)=\mathrm{GCD}(f(x), x^{p^j}-x)$.

5- We test g(x). We have two actions depending on whether g(x) is a constant or not.

5.1- g(x) is equal to a constant :

. MAJLI: procedure to update the table of degrees (no divisor of j can be a possible degree of an irreducible factor of f(x)).

. $j \leftarrow j+1$

. If $j>n_2$ THEN f(x) is irreducible : end of the program
ELSE Go back to 4.1-

5.2- g(x) is not equal to a constant :

. STIKEL: Determination of the parity s' of the number of factors of g(x).

. PEELER: Factorisation of g(x), in two steps :

.. Using s', j and the degree of g(x) we determine the set of possible degrees of irreducible factors of g(x). Solutions are deleted using indications from table of degrees updated in 5.1 by MAJLI.

.. FACTOR: Procedure to compute the irreducible factors of g(x) having the same degree. Here we **use** the Norm and Trace functions.

. $f(x) \leftarrow f(x)/g(x)$. s is updated from s'.

. IF f(x)=constant THEN end of the program
ELSE Update the Q matrix and go back to 3-

PART III : EXPERIMENTAL RESULTS

In this part we give experimental results :

3.1. COMPARISON BETWEEN #S AND m

We give a comparison between #S and m, as well as an example of maximal length sequence of squares/non-squares in GF(999983).

P	S	m	$\sqrt[4]{p}$	p	S	m	$\sqrt[4]{p}$	p	S	m	$\sqrt[4]{p}$
23	9	7	2	9973	26	25	9	80021	32	33	16
29	9	9	2	12721	27	30	10	90089	32	35	17
31	9	7	2	14779	27	27	11	99991	33	42	17
37	10	8	2	14783	27	25	11	100999	33	29	17
41	10	8	2	14983	27	27	11	129971	33	35	18
43	10	7	2	15013	27	36	11	131063	33	32	19
113	13	15	3	15083	27	27	11	131101	34	41	19
151	14	13	3	15161	27	37	11	199999	35	38	21
229	15	14	3	21001	28	42	12	250007	35	36	22
257	16	13	4	22003	28	30	12	262139	35	34	22
2741	22	25	7	22291	28	30	12	275003	36	34	22
6637	25	28	9	30011	29	35	13	299993	36	35	23
7691	25	27	9	60013	31	33	15	400009	37	38	25
8501	26	27	9	65003	31	30	15	700001	38	39	28
9001	26	27	9	70019	32	28	16	999983	39	40	31

Example of squares/non-squares sequence for p=999983

For p=999983 the following sequence of squares/non-squares of length 39 occurs twice :

        111111010000110100000110011010100010010   (1 means square)

The first sequence begins at 89101, and the second begins at 390342. Moreover one can verify that 89140 is not a square, but 390381 is a square. Figure 1 gives a graphical representation of the sets of m and $\sqrt[4]{p}$.

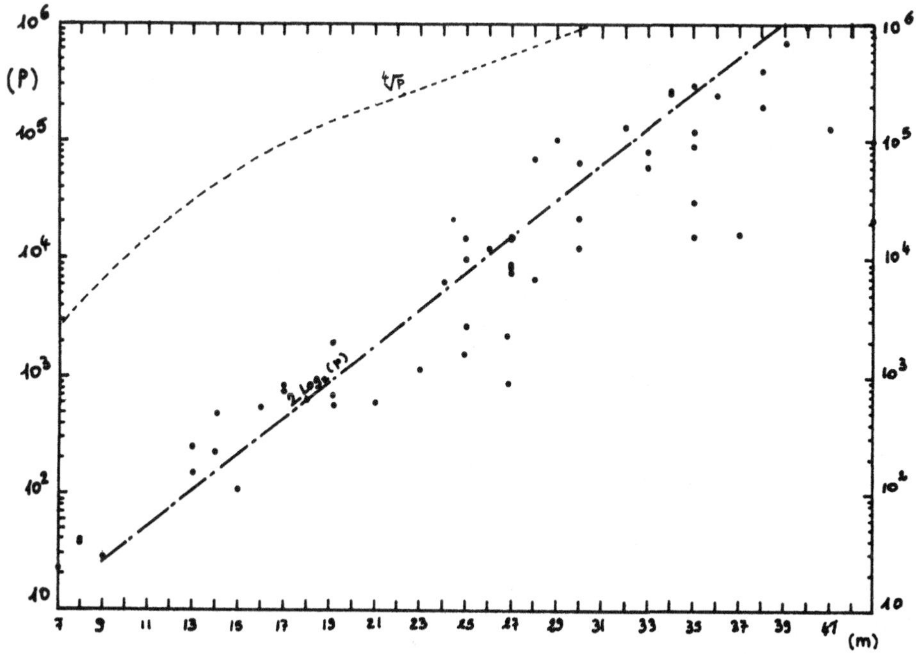

Figure 2: Length of sequences of squares/non-squares in GF(p) (p prime), according to values of the previous table

3.2. RUNNING TIMES

We now present experimental running times using FAST16.

We have compared FAST16 with MACSYMA, running on the same computer (H-Bull DPS8/70), for the same polynomials.

In the first version of this paper we used non-randomly generated polynomials. Following a remark of the referee, we give here polynomials whose coefficients are randomly generated (uniform distribution, [10] page 9).

- For degrees less than or equal to 80 we construct 5 polynomials which give rise to an average running time t_o. Among these polynomials we select the one whose running time is close to t_o, and we factorise it using MACSYMA, to obtain the corresponding running time.

- For degrees greater than 80 we only factorise one polynomial.

Remarks - There does not appear to be any significant difference between results we had with non-randomly generated polynomials, and those we have using randomly generated ones.

- The experimental results we give here are not relevant from a statistical point of view. They only give an idea of the difference between FAST16 and MACSYMA.

In the following table we give the value of p, the degree of the polynomial to be factorised, its coefficients (by increasing degrees), and the degree of the irreducible factors (put in brackets). The running times are expressed in seconds.

d	coefficients (degree of irreducible factors)	mean	FAST16	Macsyma
	p=31			
12	5 25 13 6 3 5 28 17 21 28 16 10 2 (1,4,7)	0.66	0.61	1.42
40	8 4 1 18 22 21 5 4 20 24 23 18 13 29 2 30 10 2 7 22 17 0 30 18 6 28 21 16 28 28 16 23 3 12 7 19 21 5 22 25 4 (2,3,7,11,17)	5.26	5.56	18.11
50	12 0 19 24 1 18 10 30 20 12 9 23 26 18 3 10 15 25 26 9 21 0 25 7 26 3 15 11 20 29 27 4 21 19 13 29 28 4 5 13 1 2 13 28 23 19 0 29 4 0 6 (1,5,20,24)	7.76	7.90	29.03
80	2 14 3 12 22 16 22 15 16 9 30 1 6 22 12 26 5 10 26 16 3 30 22 4 18 28 2 26 18 18 29 16 18 33 20 15 22 25 12 2 19 12 21 10 16 29 8 16 8 7 10 15 27 16 11 6 9 4 8 7 3 19 16 13 25 25 16 7 24 18 10 25 18 5 30 1 22 16 5 7 21 (3,77)	21.83	21.60	57.53
100	0 14 6 24 23 29 27 2 16 9 4 12 22 18 29 0 24 18 8 11 18 29 18 25 0 18 13 17 2 26 8 25 2 24 27 1 30 8 29 0 11 28 27 30 22 9 0 27 29 19 4 1 28 15 5 28 26 11 26 11 11 7 30 7 10 23 1 8 7 19 10 27 22 12 22 21 26 16 13 23 18 27 9 7 23 2 11 12 18 21 29 5 19 16 7 27 22 18 23 18 15 (1,35,64)	–	43.99	152.90
140	6 9 2 6 1 16 2 1 14 4 26 0 23 4 13 2 25 0 2 0 6 10 5 27 21 4 21 13 11 9 12 11 25 18 23 2 2 30 3 8 0 7 5 28 5 13 10 1 30 21 21 16 23 29 25 16 12 30 23 7 13 15 11 25 5 26 25 10 19 9 9 28 20 3 11 27 14 13 21 7 21 14 12 2 30 23 25 23 3 18 29 16 8 15 20 1 0 24 23 21 0 14 20 1 12 0 19 20 3 0 13 6 28 3 8 13 9 29 21 1 2 28 30 1 20 18 1 28 28 28 13 7 12 24 28 1 28 10 18 25 19 (1,4,10, 26,49,50)	–	108.32	340.36
	p=251			
12	77 128 136 161 87 159 230 41 200 211 6 138 236 (12)	0.86	0.84	4.26
20	68 228 44 73 237 233 58 18 23 217 33 170 206 172 165 45 195 129 42 2 178 (1,5,14)	1.62	1.47	10.43
30	206 230 242 199 2 27 56 13 35 192 206 192 174 223 162 126 231 13 103 202 51 0 99 170 83 6 20 225 229 12 90 (1,2,6,6,15)	4.52	4.60	24.43
50	31 26 18 163 87 192 78 145 138 93 43 87 49 192 112 150 45 193 131 186 190 187 132 60 53 153 25 248 201 157 54 61 144 121 194 182 191 96 168 205 47 154 40 111 26 50 242 99 148 150 7 (1,5,5,7,32)	14.37	14.05	69.70

70	62 162 141 144 155 237 159 90 209 221 102 123 126 64 84 0 181 140 53 223 174 200 213 183 174 239 158 177 0 148 227 58 24 122 71 137 109 169 44 176 93 148 118 177 39 93 111 135 184 191 62 51 83 123 13 10 30 11 157 161 18 57 222 191 129 43 161 196 40 215 180 (1,28,41)	35.64	35.11	141.49
90	52 111 38 86 185 78 53 233 248 1 214 210 107 149 181 189 72 18 1 136 215 84 221 225 74 16 110 22 30 59 158 189 110 95 121 101 84 45 27 117 30 95 89 132 72 199 206 4 153 33 147 40 120 118 190 232 90 1 105 78 122 10 221 187 147 150 160 11 205 200 226 199 65 44 156 48 115 176 67 78 156 15 128 146 235 249 50 6 225 110 93 (1,1,3,85)	-	68.34	232.21
150	147 5 99 16 179 189 180 170 38 198 82 188 21 51 165 205 57 51 237 27 81 101 226 202 222 169 236 157 30 204 183 188 161 244 144 63 81 173 158 226 131 17 160 47 98 214 138 137 196 81 175 194 121 134 33 146 60 31 182 97 52 216 210 214 32 59 14 182 239 171 231 71 3 140 134 36 200 114 161 206 76 87 131 162 82 226 199 27 115 106 133 161 169 17 126 42 50 4 50 38 55 196 155 33 3 174 139 194 49 202 9 96 212 107 41 29 42 10 151 38 85 28 134 142 27 45 170 142 99 118 194 192 88 42 84 9 98 248 143 44 45 28 23 225 6 154 194 13 20 66 134 (1,1,3,7,7,26,34,71)	-	442.74	801.10
	<u>p=79999</u>			
5	7652 22866 18752 2016 74976 55416 (1,1,3)	0.54	0.43	117.77
7	32788 61295 41233 26900 40453 36481 42823 53285 (1,6)	0.52	0.50	335.00
12	16637 67070 35207 43801 78302 10756 12047 56810 38818 25216 32863 42346 6557 (1,1,3,7)	0.90	0.90	855.93
20	60476 31792 39952 8816 3926 8216 33026 19991 44126 48517 20351 52541 17951 39566 65051 3967 31027 38242 65876 28642 13976 (1,3,4,4,8)	2.13	2.33	-
30	35338 68284 12553 47794 46569 40005 14008 39540 47124 26025 16914 42210 49378 8344 19269 20929 16959 27090 4074 43950 68488 13009 41203 1395 49468 27480 39283 42765 35398 11250 21939 (2,2,5,10,11)	4.59	4.78	-
50	28384 5789 874 20399 45038 35335 59378 40844 75518 8429 56333 72089 19699 77699 32239 10510 34328 11395 1969 3730 26783 2140 398 26875 55688 36310 76778 22570 42169 74654 14108 34690 23599 17924 20389 18359 40478 51869 42994 19330 46433 34739 19924 2099 26963 21659 9803 51794 61943 4630 32509 (16,34)	17.14	16.92	-
70	37722 39346 23971 58096 20221 30597 61471 50596 45222 63721 70846 48721 22096 4347 40846 597 13347 41847 33346 56221 46471 1846 31471 2472 67096 21222 63346 24346 24597 13722 38971 26847 33972 11847 65221 47472 3972 43722 7096 35596 76471 19347 9597 44971 74596 45597 30222 14971 26472 67471 18346 7471 2097 20596 27721 5596 28347 41221 47097 37471 50221 78721 39597 13096 3346 8097 37722 39346 23971 58096 20221 (6,15,21,28)	44.32	43.41	-
90	23476 15042 26701 3316 11926 8341 11652 7617 63376 20517 4602 41416 24702 21567 33676 37842 5277 73366 27627 24391 9477 48417 25201 76066 26052 36091 39526 12867 67501 14517 36852 29791 6327 22441 54676 4966 61276 72991 61126 267 31726 3042 11202 29442 24552 28842 4276 40617 15376 19767 40977 23791 38577 10816 36301 55216 2277 9492 37126 79891 34602 9541 33451 64066 41176 12841 2776 63991 29502 47517 68851 29041 42702 4816 68551	-	86.47	-

	43591 63901 10366 58126 16392 10602 9166 36327 39942 32802 16842 68776 17367 28002 70891 33601 (1,1,33,55)			
150	55540 48663 2920 41482 13525 57577 45606 3697 73660 77842 7191 16762 45696 22833 13675 50302 25630 922 14185 51442 73840 17362 16596 11058 2575 71152 49405 76897 16585 64417 26740 22338 50620 56782 47725 753 21306 17872 17736 9892 19015 24187 9021 5632 70375 17227 12580 36348 43885 34117 73165 34788 13671 38857 76150 60577 29481 74197 18160 52717 76690 8512 8320 38332 36925 7053 54505 8298 30561 19443 74590 30363 34221 77182 13326 68527 32031 79897 460 5542 20616 77212 40120 48532 59725 16252 44556 54622 46611 17268 35391 72187 9771 18633 8001 3352 23331 33723 26511 37117 57040 55912 38170 13732 65650 52327 41481 9697 41410 9468 25566 15888 24696 6333 37675 50677 72130 74422 71935 16818 7840 33363 18096 18307 37825 43402 21531 71647 12460 70417 43866 33963 68995 6532 26725 64252 45306 18247 14860 3393 76765 69562 23020 21633 71875 73852 47830 8598 16011 28867 69040 (1,1,9,10 11,14,21,22,26,35)	–	419.77	–
	<u>p=100999</u>			
3	4951 23191 18301 22591 (1,2)	0.35	0.32	108.42
4	54170 13982 53525 58202 30605 (1,1,2)	0.46	0.39	102.84
5	87202 47230 16698 6786 39967 48190 (2,3)	0.54	0.40	317.13
7	43532 43475 28127 1430 10247 14735 24767 9140 (1,2,4)	0.54	0.52	469.79
10	14198 99660 9593 51315 10137 22571 11832 28425 10178 12506 73047 (1,1,2,6)	0.83	0.90	746.84
15	61568 38129 39008 71414 67748 74774 60038 34709 23753 22094 79268 85304 15083 2714 35948 64574(1,1,1,2,2,8)	1.37	1.52	1526.26
20	64189 94723 23925 85888 74734 87853 18244 7243 20704 43183 62989 23548 19099 7089 33034 40303 24045 2443 23130 8008 22414 (1,3,4,12)	2.17	2.13	3595.20
30	60947 13910 92 5191 887 54695 53207 27050 12178 16505 19048 22186 69317 47840 12988 3095 97682 4700 32777 34655 16022 58085 20668 96365 24587 69620 40907 5101 31127 13681 80747 (1,1,2,26)	5.02	4.45	-
50	24208 12964 56323 46699 63238 8884 47203 49144 61093 24855 17409 765 18898 14250 1314 18060 72778 37819 28168 4530 4483 27690 71473 88174 57013 15735 19854 99619 6493 57079 25558 76864 62173 21724 45463 30034 56428 54544 57943 19879 18508 30664 2499 22774 22914 5460 37003 26344 2518 55804 22708 (1,1,1,2,3,3,3,5, 9,10,12)	17.10	17.33	-
70	4605 65533 14014 5398 22125 43813 9060 20154 85069 65293 51529 35983 24064 35473 22924 15639 6484 34228 96244 4494 42454 23308 56614 9298 63349 31963 37909 26304 45544 5319 67129 55633 5040 5623 47524 87913 41209 96628 45469 45268 52429 20709 71089 18823 54199 87613 55759 37828 51019 34843 4230 19033 69889 77023 77749 8814 14059 18654 28444 30043 79279 12234 29314 39598 16924 19513 23235 15853 17119 25494 58954 (1, 2,4,7,26,30)	41.00	38.64	-
199	86381 32711 24832 60011 95531 5437 6982 20111 90056 48536 69881 56711 25207 5512 39656 34811 23981 83486	-	804.68	-

	56681 50036 25507 91961 98456 29261 34406 30686 29981 51236 68306 68411 76256 70961 81956 53261 34781 77186 2482 7987 12682 2411 42881 72461 9206 88511 7031 28312 98231 3862 18682 99536 54506 90836 88331 67511 18907 23936 98606 21986 63806 27911 99881 24836 54956 69011 47156 59186 70856 74111 58556 23786 4882 20962 66581 87386 25282 38186 54356 98111 58931 70286 50006 50336 10957 21386 92906 39686 9982 3986 31181 21412 44756 46211 16957 17512 3532 58061 60731 11362 14681 17036 45131 92711 62081 46886 20531 20437 27356 12862 71306 52286 17381 15461 56831 42761 26531 89186 38981 2861 21056 31286 881 39461 57206 89261 100031 17561 84356 66236 16057 32786 57506 74711 29456 12011 94781 13436 90356 33986 27682 51161 7256 53711 12956 36011 95156 59936 34481 91736 44681 13537 2257 55211 69581 71261 67406 11062 29231 87611 50681 82286 13882 73586 19331 50261 50906 6686 29606 4736 23182 10661 30881 7586 14332 51761 6532 41936 1856 56861 17932 6536 36881 3712 25957 70136 57281 20936 13732 80861 18307 53036 9382 33086 42956 4136 23906 22436 41981 15112 91556 4162 (1,4,17,48,129)			

CONCLUSION

We gave a simple scheme to represent factorisation of polynomials. Classical algorithms can be viewed very easily through this scheme. The main idea to factorise is to deduce one element in $GF(p^s)$ from one another which is in $GF(p^r)$ (with s strictly less than r).

We gave experimental results : FAST16 running times comparatively with those obtained by MACSYMA, sequences of squares/non-squares in GF(p). The result of Burgess ($O(\sqrt[4]{p})$) is better than that of H. Davenport and J. Erdos ($O(\sqrt{p})$). It is possible that m is equivalent to : $\log_2(\log_2 p)\sqrt[4]{p}$, but other experiments are necessary with greater values of p.

From a general point of view, software programs for polynomial factorisation can be improved if a better method is given (and then the complexity will decrease), or if better management of data is achieved. As J.E. Hopcroft and R.E. Tarjan say: an efficient program must be realized using some "intelligent" data management.

Acknowledgements

We thank the referee who helped us to clarify several points of the previous version of this paper, and who gave us the reference of R.E. Tarjan and J.E. Hopcroft [17].

We thank Professor D. Lazard (Greco "Calcul Formel") for his financial support for our experiments with Macsyma.

REFERENCES

(1) BERLEKAMP E.R.
"Algebraic coding theory" MacGraw Hill, New York (1968)

(2) BERLEKAMP E.R.
"Factoring polynomials over large finite fields"
Math. Comp. Vol 24, pp 713-735 (1970)

(3) BURGESS D.A.
"The distribution of quadratic residues and non-residues"
Mathematika Vol 4, 106-112 (1957)

(4) CAMION P.
"Un algorithme de construction des idempotents primitifs d'idéaux sur $\mathbb{F}_q$"
Annals of Discrete Math. Vol 12,55-63 (1982)

(5) CAMION P.
"A deterministic algorithm for factorising polynomials of $\mathbb{F}_q[X]$"
Annals of Discrete Math. vol 17, 149-157 (1983)

(6) CANTOR D.G. and ZASSENHAUS H.
"A new algorithm for factorising polynomials over finite fields"
Math. Comp. vol 36 154-163 (1981)

(7) DAVENPORT H., ERDOS P.
"The distribution of quadratic and higher residues"
Publ. Math. Debrecen vol 2 252-265 (1952)

(8) GOLOMB S.W., WELCH L.R., HALES A.
"On the factorisation of trinomials over GF(2)"
J.P.L. Memo 20-189 (July 1959)

(9) HUDSON R.H.
"On the sequences of consecutive quadratic non-residues"
J. of Number Theory, Vol 3 178-181 (1971)

(10) KNUTH D.E.
"The Art of computer programming", Vol 2 Seminumerical algorithms
Addison-Wesley Publishing Company (2th Edition)

(11) MacWILLIAMS F.J.
"The structure and properties of binary cyclic alphabets"
Bell System Technical J. Vol 44 303-332 (1965)

(12) MOENCK R.T.
"On the efficiency of algorithms for polynomial factoring"
Mathematics of computation, Vol 31 235-250 (1977)

(13) POLI A., RIGONI C.
"Experimental comparison between softwares for binary polynomial factorisation" (In French)
Revue Traitement du Signal Vol 1n°2-2 211-215 (1984)

(14) POLI A.
"Construction of primitive idempotents for n variable codes"
LNCS/Springer Verlag, n°228, 25-35 (1985)

(15) RABIN M.O.
"Probabilistic algorithms in finite fields" MIT/LCS/TR.213 (1979)

(16) TARJAN R.E., HOPCROFT J.E.
"An interview with the 1986 A.M. Turing award recipients"
Communications of the ACM vol 30 n°3 214-222 (1987)

(17) ZASSENHAUS H.J. "On Hensel factorisation" J. Numb. Th. 291-311(1969)

CHARACTERIZATION OF COMPLETELY REGULAR CODES THROUGH P-POLYNOMIAL ASSOCIATION SCHEMES.

by

Josep Rifà & Llorenç Huguet
Dpt. d'Informàtica. Fac. Ciències
Universitat Autònoma de Barcelona
Catalunya

Abstract:

Given a projective linear code C(n,k) with s nonzero weigths we know that if $C^{\perp}$ is a uniformly packed code, or more generally if $C^{\perp}$ is a completely regular code, then the covering radius ρ of $C^{\perp}$ is equal to s. Furthermore, the restriction of the Hamming scheme H(n,q) to C is an association scheme and its dual association scheme is isomorphic to its Ω-scheme, an association scheme defined on the columns of the generator matrix of C.

On the other hand there exist projective linear codes C whose restriction of H(n,q) is an association scheme although its orthogonal code $C^{\perp}$ is not completely regular. In other words, there exist projective linear codes C with ρ = s but $C^{\perp}$ is not completely regular.

In this paper we present two main results:

1.- $C^{\perp}$ is a completely regular code if and only if ρ = s and the restriction of H(n,q) on C is an association scheme.

2.- $C^{\perp}$ is a completely regular code if and only if the Ω-scheme is a P-polynomial association scheme.

I.-**Definitions and background.**

Let $F=GF(q)$ be the finite Galois field of $q = p^r$ elements, p being a prime.

Let C be a linear code (n,k), that is a k-dimensional vector subspace of F^n. We denote by G the generator matrix and by s the number of nonzero weights of C.

I.1.- **Definition**

Let $\Omega = \{\Omega_1, \Omega_2, \ldots, \Omega_n\}$ be the set of columns of G, also called the coordinate forms of C. Let C' be the vector space on F generated by Ω. We will call it the **dual code** of C.

C is a **projective code** if and only if $a \cdot \Omega \cap \Omega = \emptyset$, for all $a \in F-\{1\}$.

I.2.- **Definition**

Given a linear code C(n,k) on F, their orthogonal code is

$$C^{\perp} = \{v \in F^n \mid v.c = 0 \text{ for all } c \in C\}$$

where v.c is the componentwise dot product.

I.3.-**Definition**

For all $x \in F^n$ and any natural number a $(0 \leq a \leq n)$ we define:

$$B(x,a) = \#\{ c \in C^{\perp} \mid d_H(x,c)=a \}$$

where d_H is the Hamming distance.

$C^{\perp}$ is a **completely regular code** if for any $x \in F^n$ and for any a $(0 \leq a \leq n)$ then the value of B(x,a) only depends on the minimum Hamming distance $d_{min}(x, C^{\perp})$.

Remark: B(x,a) is precisely the number of vectors with weight a in the coset $x + C^{\perp}$.

I.4.-**Definition**

Set $\rho(x) = \min_{0 \leq a \leq n} \{ a \mid B(x,a) \neq 0 \}$

The **covering radius** of $C^{\perp}$ is defined as:

$$\rho = \mathrm{Max}_{x \in F^n} \{ \rho(x) \}$$

I.5.-**Notations**

Let $\mathbb{C}$ be the set of complex numbers. If H is a vector space on F then the set of all applications from H to $\mathbb{C}$ is a vector space on $\mathbb{C}$. It will be denoted by $\mathbb{C}[H]$.

We consider $\{f_u : h \in H\}$ as a canonical basis of $\mathbb{C}[H]$, $\{f_u\}$ being the set of characteristic functions of the points of H.

I.6.-**Definition**

Let H be a vector space on F. Let $\{G_i\}_{i=o,\ldots,d}$ be a partition on H whose equivalence classes satisfie that if $x \in G_i$ then $-x \in G_i$.

The pair $(H,\{G_i\})$ is called a **scheme of d classes on H.**

I.7.-**Definition**

Given two schemes $(H,\{G_i\})$ and $(Z,\{M_i\})$, with the same number of classes, we will say that $\Phi : H \longrightarrow Z$ is an **isomorphism of schemes** if it is an isomorphism of vector spaces between H and Z and for all class C_i there exist some class M_j such that $h(G_i) = M_j$.

I.8.-**Definition**

Let $(H,\{G_i\})_{i=o,\ldots,d}$ be a scheme of d classes on H and let $g_i \in \mathrm{End}_{\mathbb{C}}\mathbb{C}[H]$ be the adjacency endomorphism associated to G_i, that is:

$$g_i(f_x,f_y) = \begin{cases} 1 & \text{if } x-y \in G_i \\ 0 & \text{otherwise.} \end{cases}$$

If the adjacency endomorphisms $\{g_i\}$ generate a (d+1)-dimensional commutative subalgebra of $End_{Ç}Ç[H]$ then $(H,\{G_i\})$ is an **association scheme of d classes on H** (invariant under translations, commutative and symmetric).

I.9.-**The Hamming scheme H(n,q)**

Considering the n-dimensional vector space F^n we could define the partition $\{R_i\}$ as

$$R_i = \{ x \in F^n : W_H(x) = i \} \quad \text{for } i=0,1,\ldots,n$$

where $W_H(x)$ is the Hamming weight, i.e. the number of nonzero coordinates in x.

The pair $(F^n,\{R_i\})$ is an **association scheme of** n **classes** called the **Hamming scheme** on F^n and denoted by H(n,q).

I.10.-**Definition**

Given an association scheme $(X,\{R_i\})$ with d classes (commutative and symmetric), we define the application:

$$d: X\times X \text{ -------> } N=\{0,1,2,\ldots,d\}$$

$$(a,b) \text{ ------> } d(a,b) = k \text{ if and only if } (a,b) \in R_k$$

$(X,\{R_i\})$ is a **metric** association scheme if d is a distance on X such that for all a,b ∈ X with d(a,b) = k, there exist at least one point c ∈ X which d(x,c) = 1 and d(x,b) = k-1.

I.11.-**Definition**

Given an association scheme $(X,\{R_i\})$ with d classes (commutative and symmetric) it is **P-polynomial** if there exists polynomials with real coefficients $m_i(x)$ of degree i (i=0,1,2..,d) such that $D_i = m_i(D_1)$ where D_i are the adjacency matrices associated to R_i.

I.12.-**Theorem** ([1] theorem 5.6)

A symmetric association scheme is P-polynomial if and only if it is metric.

II.- Schemes on C and $C^{\perp}$:

In this section we give some schemes defined on C and $C^{\perp}$ considered as k-dimensional and (n-k)-dimensional vector subspaces of F^n, respectively.

Recall that s is the number of nonzero weights in C, ρ is the covering radius of $C^{\perp}$ and Ω is the set of coordinate forms of C.

II.1.- The pair $(C,\{D_i\})$, where $x \in D_i$ if and only if $W_H(x) = i$, is a scheme of s classes on C, is the restriction of H(n,q) to C.
Remarks: Let $\{d_i \in End_{Ç}Ç[C]\}$ be the adjacency endomorphisms associated to the partition $\{D_i\}$.

There exist a basis of $Ç[F^k]$ which diagonalizes all d_i. The elements in the basis are the characters of $\hat{}C = \{ \sigma_a \mid a \in C \}$:

$$\sigma_a \ : \ \begin{array}{l} C \text{ ------> } Ç \\ b \text{ ------> } \sigma_a(b) = w^{Tr.\emptyset(a,b)} \end{array}$$

where w is a p-th root of unity, $\emptyset$ is the metric in F^n and Tr. is the trace application from GF(q) to GF(p).

II.2.- The pair $(\hat{}C,\{\hat{}D_i\})$, where "$\sigma_a \approx \sigma_b$ if and only if the eigenvalues associated to eigenvectors σ_a and σ_b are the same on all D_i", is a scheme of r classes on $\hat{}C$, called the **dual scheme** of (C,{D>i<}), (Always $r \geq s$).

II.3- The pair $(F^n/C^{\perp}, \{P_i\})$, where

"two cosets of $C^{\perp}$ are in the same class P_i if and only if they have the same weight enumerator polynomial",

is a scheme of r classes on F^n.

II.4.- The pair $(F^n/C^{\perp}, \{S_i\})$, where

"the coset $x + C^{\perp} \in S_i$ if and only if the minimum weight of the coset is i" $(i = \min\{W_H(x+y) \mid y \in C^{\perp}\})$,

is a scheme of ρ classes on $F^n/C^{\perp}$.

II.5.- Set $\Omega' = F \cdot \Omega$. Taking the dual code C' of C ($\Omega' \subset C'$), the pair $(C', \{G_i\})$, where

"$a \in G_i$ if and only if $i = \min\{ k \mid a$=addition of k elements of $\Omega'\}$",

is a scheme of ρ classes on C' that we will call the **Ω-scheme associated to C**.

III.-**Main results:**

In this section we characterize when the schemes defined above are association schemes. Moreover we give the necessary and sufficient conditions for the existence of isomorphims between pairs of them. Therefore we can characterize the codes such that their orthogonal $C^{\perp}$ is completely regular in terms of association schemes defined on C.

III.1.- We can consider the diagram

$$
\begin{array}{ccccc}
(\hat{}C, \hat{}D_i) & \xrightarrow{f} & (F^n/C^{\perp}, P_i) & \xrightarrow{g} & (F^n/C^{\perp}, S_i) \\
 & & & & \Big| \, h \\
 & & & & (C', G_i)
\end{array}
$$

and establish the following trivial results:

III.1.1.- $f: F^n/C^{\perp} \longrightarrow \hat{}C$
$a+c^{\perp} \longrightarrow \sigma_a$

is an algebraic isomorphism and it is an isomorphism of schemes, too ([1],theorem 6.9).

III.1.2.- $h: F^n/C^{\perp} \longrightarrow C'$
$a+c^{\perp} \longrightarrow a.$ $(= i\emptyset.a/C)$,

where i is the inner contraction of $\emptyset$. The h is an algebraic isomorphism and it is an isomorphism of schemes, too.

III.1.3.- $g: F^n/C^{\perp} \longrightarrow F^n/C^{\perp}$
$a+c^{\perp} \longrightarrow a+c^{\perp}$

is only an algebraic isomorphism (g = Id), but it is not an isomorphism of schemes.

Note: g will be an isomorphism of schemes if and only if $r = \rho$, that is, if and only if $C^{\perp}$ is a completely regular code (see I.3).

III.2.-**Proposition** ([1] theorem 6.10)

The restriction of $H(n,q)$ to C, i.e. the scheme $(C,\{D_i\})$, is an association scheme if and only if $r = s$.

III.3.-**Proposition**

If $C^{\perp}$ is a completely regular code then $\rho = s$, and $(C,\{D_i\})$ is an association scheme.

Proof:

If $C^{\perp}$ is a completely regular code then $(C,\{D_i\})$ is an association scheme and g is an isomorphism of schemes, therefore $r = \rho$ (see III.1.3).

Because always $r \geq s$ and $s \geq \rho$ then we have $r = \rho = s$.

Remark: With these conditions the composition between f, g and h is an isomorphism of schemes. Then the Ω-scheme of C is an association scheme isomorphic to the dual scheme of the restriction of H(n,q) to C.

III.4.-Observations

III.4.1.- There are linear codes C such that $\rho = s$ but $C^{\perp}$ is not a completely regular code. For example, the code with generating matrix:

$$\begin{bmatrix} 1&0&1&0&0&0&0&0&0&1&0&1 \\ 0&1&1&0&0&0&0&0&0&0&1&1 \\ 0&0&0&1&0&1&0&0&0&0&1&1 \\ 0&0&0&0&1&1&0&0&0&1&1&0 \\ 0&0&0&0&0&0&1&0&1&1&1&0 \\ 0&0&0&0&0&0&0&1&1&1&0&1 \end{bmatrix}$$

III.4.2.- Also there are some linear codes C such that the restriction of H(n,q) is an association scheme but $C^{\perp}$ is not a completely regular code. For example, the code with generating matrix:

$$\begin{bmatrix} 1&0&0&1 \\ 1&1&1&0 \\ 0&0&1&1 \end{bmatrix}$$

III.5.-**Theorem**:

Let C(n,k) be a projective linear code such that $\rho = s$. If the restriction of H(n,q) to C is an association scheme then $C^{\perp}$ is a completely regular code.

Proof:

Obviously, from III.2 we have r = s and it is assumed that $\rho = s$; consequently $\rho = r$. The result III.1.3 ensures that $C^{\perp}$ is completely regular.

III.6.-**Notation and remarks**

Set $H = F^n/C^{\perp}$ and let $\{g_i \in End_{\mathbb{C}}\mathbb{C}[H]\}$ be the associated adjacency endomorphisms as in II.5. We define the endomorphisms $\{z_i \in End_{\mathbb{C}}\mathbb{C}[H]\}$, referred to the canonical basis of $\mathbb{C}[H]$ (see I.5), as follows:

$$z_i(f_u, f_v) = B(v-u, i)$$

There exist a basis of $\mathbb{C}[H]$ which diagonalizes simultaneously all endomorphisms g_i and z_i. The elements of these basis are $\Sigma_{u \in H}\, \sigma(u) f_u$ where $e \in \hat{}H$. Moreover, the eigenvalues of z_i are the differents values of $P_i(W_j)$, where P_i are the Krawtchouk polynomials and W_j are the distincts nonzero weights of C.

III.7.-**Proposition**

From our previous notations we stablish that:

a): $z_o = g_o$; $z_1 = g_1$,

b): $\{z_o, z_1, \ldots, z_n\}$ generates a s-dimensional algebra isomorphic to the algebra generated by g_o and g_1, and

c): Let $A(\{z_i\})$ be the algebra generated by $z_o, z_1, \ldots, z_n$. This $A(\{z_i\})$ is contained in the algebra $A(\{g_i\})$ generated by $g_o, g_1, \ldots, g_{\rho}$.

Proof:

a): By definition of the endomorphisms z_i and g_i.

b): From III.5 we can write each z_i as a polynomial, of degree i, in the variable z_1, then by a) we achieve the proof.

c): Trivially from a) and b).

III.8.-**Proposition:**

If $C^\perp$ is a completely regular code then the Ω-scheme of C is an association scheme which is isomorphic to the restriction of H(n,q) to C.

Proof:

From III.2 and III.3 the restriction of H(n,q) to C is an association scheme. Moreover, from III.1.1, III.1.2 and III.1.3 and knowing that the composition of f, g and h is an isomorpfism of schemes, we can conclude that the dual association scheme $(\hat{}C,\{\hat{}D_i\})$ is isomorphic to the Ω-scheme.

III.9.-**Theorem:**

If the Ω-scheme is an association scheme then $C^\perp$ is a completely regular code.

Proof:

If the Ω-scheme is an association scheme then the algebra $A(\{z_i\}$ has dimension $\rho+1$. (see I.8).

From III.7 we can write $s \leq \rho$ and therefore $\rho = s$. In this situation we have $A(\{z_i\}) = A(\{g_i\})$ and each z_i can be considered as a linear combination of $g_0, g_1, \ldots, g_\rho$, that is to say, $C^\perp$ is a completely regular code.

III.10.-**Proposition**:

If the Ω-scheme is an association scheme then it is P-polynomial.

Proof:

We define $d: C'xC' \text{------>} N$
$(x,y) \text{------>} d(x,y) = k$ if and only if $y-x \in G_k$

It is not difficult to prove that the Ω-scheme is a metric scheme and thrrefore it is P-polynomial (see theorem I.12).

REFERENCES:

[1] DELSARTE P.: "An Algebraic Approach to the Association Schemes of Coding Theory", Philips Research Rep. Supp. 10 (1975).

[2] DELSARTE P.: "Four fundamental parameters of a code and their combinatorial significance", Information and control 23, 407-438. (1973).

[3] GOETHALS J.M. & VAN TILBORG H.C.A.: "Uniformly packed codes" Philips Research Reports 30 , 9-36. (1975).

[4] RIFÀ J.: "Equivalències entre estructures combinatòricament regulars: Esquemes, grafs i codis". Tesi doctoral. Universitat Autònoma de Barcelona. 1987.

CODES ON HERMITIAN CURVES

Mohammad Amin Shokrollahi
Fakultät für Informatik
Universität Karlsruhe
D - 7500 Karlsruhe 1

1. Introduction

A few years ago V. D. Goppa introduced a certain class of codes related to algebraic curves over the finite field $GF(q)$. The construction of these codes is as follows :

Let **X** be an algebraic curve of genus g over $GF(q)$ and let $Q, P_1, P_2, \ldots, P_n$ be its $GF(q)$-rational points. Denote by $\Omega(\alpha Q - \sum P_i)$ the $GF(q)$-vector space of differentials on **X** vanishing with multiplicity of at least α at Q, regular everywhere outside P_i, and having poles of at most the first order at P_i. Define the mapping $\varphi : \Omega(\alpha Q - \sum P_i) \longrightarrow GF(q)^n$ by $\omega \longmapsto (Res_{P_1}(\omega), \ldots, Res_{P_n}(\omega))$. The image of φ is an (n, k, d)-code C. Assume that $2g - 2 < \alpha < n$. Then, by the Riemann - Roch theorem, we have

$$\begin{aligned} k &= n - \alpha + g - 1 \\ d &\geq \alpha - 2g + 2. \end{aligned} \tag{1}$$

Let $L(\alpha Q)$ denote the $GF(q)$-vector space of functions f on **X** (i.e. in the function field of **X**) which have a pole of the order of at most α at Q. We define a new code C' by

$$C' = \{(f(P_1), \ldots, f(P_n)) \mid f \in L(\alpha Q)\}.$$

In fact, C' is the image of $\varphi' : L(\alpha Q) \longrightarrow GF(q)^n$ defined by $f \longmapsto (f(P_1), \ldots, f(P_n))$. The dimension of C' equals that of $L(\alpha Q)$ if φ' is injective. Now $\ker \varphi' = L(\alpha Q - \sum P_i)$, so $\ker \varphi' = \{0\}$ since $\alpha < n$. Thus

$$\dim C' = \dim L(\alpha Q) = \alpha - g + 1 = n - \dim C.$$

Now, let $\omega \in \Omega(\alpha Q - \sum P_i)$ and $f \in L(\alpha Q)$. Put $\underline{c} := \varphi(\omega)$ and $\underline{c}' := \varphi'(f)$. Then

$$\sum_{i=1}^{n} c_i c_i' = \sum_{i=1}^{n} f(P_i)\, Res_{P_i}(\omega) = \sum_{i=1}^{n} Res_{P_i}(f\omega) = 0,$$

by the residue theorem. Thus, $C' = C^{\perp}$.

The above observation gives an efficient method for calculating the parity check matrix H_C of C whenever a basis $\{ f_1, f_2, \ldots, f_r \}$ of $L(\alpha Q)$ is known. One has :

$$H_C = \begin{pmatrix} f_1(P_1) & f_1(P_2) & \ldots & f_1(P_n) \\ f_2(P_1) & f_2(P_2) & \ldots & f_2(P_n) \\ \vdots & \vdots & \ddots & \vdots \\ f_r(P_1) & f_r(P_2) & \ldots & f_r(P_n) \end{pmatrix}.$$

In the following we want to investigate a certain class of curves over an infinite tower of constant fields, and give a method for the computation of H_C .

2. Hermitian curves : Basic facts

Definition. Let $F = GF(p^{2h})$, $h \in \mathbf{N}$ and $r = p^h + 1$. The algebraic function field $K := F(x, y)$ defined by $x^r + y^r + 1 = 0$ is called the *unitary Fermat function field*, and the projective curve $\mathbf{X} : x^r + y^r + z^r = 0$ is called the *Hermitian curve.*

Lemma 1. *Let* $\mathbf{X}$ *be a Hermitian curve over* $GF(p^{2h})$. *Then :*
(i) The genus $g(\mathbf{X})$ *of* $\mathbf{X}$ *satisfies* $g(\mathbf{X}) = (p^{2h} - p^h)/2$.
(ii) The number of rational points of $\mathbf{X}$ *over* $GF(p^{2h})$ *is :* $\overline{n}_{GF(p^{2h})}(\mathbf{X}) = p^{3h} + 1$.

Proof . (i) $\mathbf{X}$ has no singularities, so

$$g(\mathbf{X}) = \frac{1}{2}(r-1)(r-2) = \frac{p^{2h} - p^h}{2}.$$

(ii) Let $z = 0$. Then we may take $y = 1$ and try to solve $x^r + 1 = 0$. If $p \neq 2$ we can solve $x^r = -1$ because $GF(p^{2h})$ contains a primitive $2r$-th root of unity. For $p = 2$ we need to solve $x^r = 1$ which is also possible. So, for $z = 0$, we get r rational points of $\mathbf{X}$. Now, let $z = 1$. We have to solve the equation $x^r + y^r + 1 = 0$ for $x, y \in GF(p^{2h})$. Let x be fixed. If $x^r + 1 = 0$, there is exactly one solution for y. In all other cases we can find r solutions for y ; in fact, let x be fixed and $x^r + 1 \neq 0$. Then $x^r \in GF(p^h)^\times$, hence $x^r + 1 \in GF(p^h)^\times$. It follows that $r \,|\, ord(x^r + 1)$, which implies that $y^r = -(x^r + 1)$ has r solutions for y. So $\mathbf{X}$ has altogether

$$r + r + r(p^{2h} - r) = 2(p^h + 1) + (p^h + 1)(p^{2h} - p^h - 1) = p^{3h} + 1$$

$GF(p^{2h})$-rational points.

Q.E.D.

Our next task is to introduce a connection between the rational points of the Hermitian curve $\mathbf{X}$ and the set $\mathbf{P}_1(K/F)$ of the prime divisors of K/F of degree 1.

To this end, consider the rational function field $F(x)$. Let

$$(x-\varsigma_i)\simeq\frac{\wp_i}{\wp_\infty},\qquad \wp_i,\wp_\infty\in\mathbf{P}(F(x)/F),\qquad i=1,\dots,r\,,$$

where ς_i satisfy $\varsigma_i{}^r+1=0$. The $\wp_i$ and $\wp_\infty$ are prime divisors of degree 1 since $d(\wp_i)=d(\wp_\infty)=\big(F(x-\varsigma_i):F(x)\big)=1$. In K, let $y=P_0/P_\infty$. This yields the divisor equality

$$\frac{P_0{}^r}{P_\infty{}^r}=\frac{\prod_{i=1}^r\wp_i}{\wp_\infty{}^r}$$

which can only be fullfilled if all the $\wp_i$'s are completely ramified in k that is, if there exist $P_i\in\mathbf{P}(K/F)$ such that $\wp_i=P_i{}^r$ for all $i=1,\dots,r$. So

$$(y)\simeq\frac{\prod_{i=1}^r P_i}{P_\infty}.$$

What about $\wp_\infty$? Consider the field $F(1/x)$. $\tilde{y}:=y/x$ is a primitive element of K over $F(1/x)$ and has the minimal polynomial

$$f(Y,x)=Y^r+\big(1+x^{-r}\big).$$

Now, $x^{-1}\equiv 0 \bmod \wp_\infty$ and so $f(Y,x)\equiv\prod_{i=1}^r(Y-\varsigma_i)\bmod\wp_\infty$. Clearly, the factors $(Y-\varsigma_i)$ are all mutually prime, so we can apply Hensel's Lemma to see that $f(Y,x)=\prod_{i=1}^r f_i(Y,x)$ in the $\wp_\infty$-adic completion $K_{\wp_\infty}$. So, there are r extensions of the valuation $|\cdot|_{\wp_\infty}$ to K and hence $\wp_\infty$ splits completely in K :

$$\wp_\infty=\prod_{i=1}^r P_{\infty,i}.$$

Now, let $\alpha\in GF(p^{2h})$, $\alpha^r+1\neq 0$. Put $(x-\alpha)\simeq\wp_\alpha/\wp_\infty$. Then we have in K

$$\wp_\alpha=\prod_{i=1}^r P_{\alpha,i},\qquad P_{\alpha,i}\in\mathbf{P}(K/F).$$

In fact, the minimal polynomial of y over $F(x-\alpha)$ is $Y^r+\big((x-\alpha)+\alpha\big)^r+1$ and

$$Y^r+\big((x-\alpha)+\alpha\big)^r+1\equiv Y^r+(\alpha^r+1)\equiv\prod_{i=1}^r(Y-\alpha_i)\bmod\wp_\alpha.$$

Since furthermore the $(Y-\alpha_i)$ are mutually prime, we can again apply Hensel's Lemma to get the above assertion. Thus, we have proved :

Theorem 1. *Let $F=GF(p^{2h})$ and K/F be a unitary Fermat function field. Further let $\varsigma_1,\dots,\varsigma_r\in F$ such that $\varsigma_i{}^r+1=0$. Denote by $\wp_i$ the numerator divisors of $(x-\varsigma_i)$*

and by $\wp_\infty$ their common denominator divisor. For $\alpha \in F, \alpha^r + 1 \neq 0$, let $\wp_\alpha$ denote the numerator divisor of $(x - \alpha)$. Then :

(i) For all $i = 1, \dots, r$, $\wp_i$ is completely ramified in K ,i.e. there exist $P_i \in \mathbf{P}(K/F)$ such that $\wp_i = P_i^{\,r}$.

(ii) For all $\alpha \in F$, $\alpha^r + 1 \neq 0$, $\wp_\alpha$ splits completely in K , i.e. there exist $P_{\alpha,j} \in \mathbf{P}(K/F)$ such that $\wp_\alpha = \prod_{j=1}^r P_{\alpha,j}$.

(iii) $\wp_\infty$ splits completely in K, i.e. there exist $P_{\infty,j} \in \mathbf{P}(K/F)$ such that $\wp_\infty = \prod_{j=1}^r P_{\infty,j}$.

(iv) y has the following divisor :

$$(y) \simeq \frac{\prod_{j=1}^r P_i}{\prod_{j=1}^r P_{\infty,j}}.$$

It is known that there exists a bijective mapping from the set $\mathbf{P}_1(K/F)$ on the set of the F-rational points of the curve $\mathbf{X}$. In this specific case, the mapping is as follows :

Let $Z := \{\alpha \mid \alpha \in F, \alpha^r + 1 = 0\}$. Z has r elements, say $\varsigma_1, \dots, \varsigma_r$. Then the rational points $(\varsigma_i, 0, 1)$ correspond to the prime divisors P_i of above. For $\alpha \in F \setminus Z$, the rational points $(\alpha, y_1, 1), \dots, (\alpha, y_r, 1)$ correspond to the divisors $P_{\alpha,1}, \dots, P_{\alpha,r}$. Lastly, the points $(\varsigma_1, 1, 0), \dots, (\varsigma_r, 1, 0)$ correspond to the prime divisors $P_{\infty,1}, \dots, P_{\infty,r}$.

Our next question is the following : When does there exist a rational point of $\mathbf{X}$ with $GF(p)$-rational coefficients?

Lemma 2. *Let $p = 2$ or $p \equiv 1 \bmod 4$. Then $\mathbf{X}$ has a $GF(p)$-rational point.*

Proof. It suffices to show that under the above assumptions on p the set $GF(p) \cap Z$ is not empty. In fact, if ς lies in $GF(p) \cap Z$ then $(\varsigma, 0, 1)$ would be a $GF(p)$-rational point of $\mathbf{X}$. If $p = 2$ then $1 \in GF(p) \cap Z$, so assume that $p \neq 2$. Let ε be primitive in F. The group of the $2r$-th roots of unity is generated by $\theta := \varepsilon^{(p^h - 1)/2}$, thus $Z = \{\theta^u \mid u \equiv 1 \bmod 2\}$. We show that for $GF(p) \cap Z \neq \emptyset$, it is necessary that $p \equiv 1 \bmod 4$. Suppose that there exists an odd u such that $\theta^u \in GF(p)$. This implies that there exists $d \mid (p-1)$ such that $u \cdot (p^h - 1)/2 \cdot d \equiv 0 \bmod (p^{2h} - 1)$, which yields that $u \cdot (d/2) \equiv 0 \bmod (p^h + 1)$. Now, since u is odd, d has to be even, say $d = 2d'$, where $d' | (p-1)/2$. Thus, we get the necessary condition :

$$\gcd\Big(\frac{p-1}{2}, p^h + 1\Big) = d' \neq 1.$$

But $p^h + 1 \equiv 2 \bmod (p-1) \Rightarrow p^h + 1 \equiv 2 \bmod (p-1)/2$. Thus, $\gcd\big((p-1)/2, p^h + 1\big) = \gcd\big((p-1)/2, 2\big)$, and hence, $\gcd\big((p-1)/2, 2\big) \neq 1$ iff $p \equiv 1 \bmod 4$. Now, the condition $p \equiv 1 \bmod 4$ is also sufficient, because then $\gcd\big((p^h + 1), (p-1)/2\big) = 2$ and one can find odd u such that $2u \equiv 0 \bmod \big(p^h + 1\big)$. Then, $\theta^u \in GF(p) \cap Z$, by the above arguments.

Q.E.D.

From now on, we assume that $p = 2$ or $p \equiv 1 \bmod 4$. Furthermore we define the divisor Q by $Q := (\varsigma, 0, 1)$ where $\varsigma \in GF(p) \cap Z$.

3. Construction of parity check matrices of Hermitian Goppa codes

We have already seen that the construction of the parity check martix of a Goppa code on the algebraic curve $\mathbf{X}$ over $GF(q)$ is equivalent to finding the $GF(q)$-rational points of $\mathbf{X}$ and a basis of the space $L(\alpha Q)$ as defined in section 1.

Let $\mathbf{X}$ be the Hermitian curve $x^r + y^r + z^r = 0$ over $GF(p^{2h})$, where $r = p^h + 1$, and let $p \equiv 1 \bmod 4$ or $p = 2$. Further, let $\varsigma \in GF(p) \cap Z$, i.e. $\varsigma^r + 1 = 0$. Put $(x - \varsigma) \simeq \wp/\wp_\infty$, where $\wp, \wp_\infty \in \mathbf{P}_1(F(x)/F)$. By theorem 1, $\wp$ is completely ramified in K, i.e. $\wp = \mathcal{Q}^r$ where $\mathcal{Q} \in \mathbf{P}(K/F)$. Let the point Q correspond to $\mathcal{Q}$ (cf. section 2), i.e. $Q = (\varsigma, 0, 1)$. Then $L(\alpha Q)$ is the set of all functions $f \in K$, satisfying $ord_{\mathcal{Q}}(f) \geq -\alpha$. The following theorem permits the identification of a special basis of $L(\alpha Q)$ with a set of two-tuples of natural numbers :

Theorem 2. *Let $\mathbf{X}$ be the Hermitian curve over $F := GF(p^{2h})$ and let Q be defined as above. Then the set*

$$\left\{ \frac{y^v z^{u-v}}{(x - \varsigma z)^u} \;\middle|\; 0 \leq v \leq \min(u, r-1)\,,\, 0 \leq ur - v \leq \alpha \right\}$$

is a basis of $L(\alpha Q)$.

Proof. Following the notation of section 2, we put $P_\infty := \prod_{i=1}^r P_{\infty,i}$. Then $(x - \varsigma) \simeq \mathcal{Q}^r / P_\infty$ and $(y) \simeq P_1, \ldots, P_r \mathcal{Q} / P_\infty$. Let $(a,b) \in \{\,(u,v) \mid 0 \leq v \leq \min(u, r-1)\,,\, 0 \leq ur - v \leq \alpha\}$. Then

$$ord_{\mathcal{Q}}\left(\frac{y^b}{(x-\varsigma)^a}\right) = b - ar \geq -\alpha;$$

thus, after homogenizing the function in the parantheses, we get $y^b z^{a-b}/(x - \varsigma z)^a \in L(\alpha Q)$. Now, for $(a,b), (a',b') \in \{\,(u,v) \mid 0 \leq v \leq \min(u, r-1)\,,\, 0 \leq ur - v \leq \alpha\}$ we have

$$ord_{\mathcal{Q}}\left(\frac{y^b}{(x-\varsigma)^a}\right) = ord_{\mathcal{Q}}\left(\frac{y^{b'}}{(x-\varsigma)^{a'}}\right) \Longleftrightarrow (a,b) = (a',b');$$

in fact, the two $\mathcal{Q}$-orders are equal iff $b' - a'r = b - ar$, i.e. iff $b' - b = (a' - a)r$, which implies that $r | (b - b')$ and hence $b = b'$. Because $b' - a'r = b - ar$, the latter equality implies $(a,b) = (a',b')$. Hence, the functions $(y^b/(x-\varsigma)^a)$ have different $\mathcal{Q}$-orders as (a,b) runs over $\{\,(u,v) \mid 0 \leq v \leq \min(u, r-1)\,,\, 0 \leq ur - v \leq \alpha\}$. Therefore, they are linearly independent over $GF(p^{2h})$. So, it suffices to prove that

$$\left| \left\{ (u,v) \;\middle|\; 0 \leq v \leq \min(u, r-1)\,,\, 0 \leq ur - v \leq \alpha \right\} \right| = \dim L(\alpha Q) =: l(\alpha Q). \qquad (*)$$

The condition $0 \leq v \leq \min(u, r-1)$ implies that for $(u,v) \neq (u',v')$, the inequality $ur - v \neq u'r - v'$ is satisfied. So, in order to find the cardinality of $\{\,(u,v) \mid 0 \leq v \leq \min(u, r-1)\,,\, 0 \leq ur - v \leq \alpha\}$, it suffices to compute the cardinality of the set

$$\left\{ k \;\middle|\; 0 \leq k \leq \alpha,\, \exists\, a, b \in \mathbf{N}\,,\, 0 \leq b \leq \min(a, r-1) : k = ar - b \right\}.$$

Now note that $\{(u,v) \mid 0 \le v \le \min(u,r-1)\,,\, 0 \le ur - v \le \alpha\}$ can be decomposed into the two sets $\{(u,v) \mid 0 \le v \le \min(u,r-1)\,,\, 0 \le ur - v \le 2g-1\}$ and $\{(u,v) \mid 0 \le v \le \min(u,r-1)\,,\, 2g \le ur - v \le \alpha\}$, where $g = g(\mathbf{X})$. Let us consider the second set first :
We claim that for all $2g \le k \le \alpha$, there exist $a, b \in \mathbf{N}$ such that $0 \le b \le \min(a, r-1)$ and $k = ar - b$. Clearly, this would imply that the cardinality of the second set is $\alpha - 2g + 1$. To prove the above assertion, we apply Euclidean algorithm to find $a, b \in \mathbf{N}$ such that $k = ar - b$ and $b \le r - 1$. We only have to show that $b \le a$. To this end, note that by (1) $2g = r(r-2) - (r-2)$ and $2g \le ar - b$, since (a,b) is assumed to be an element of the second set. Hence, if $a = r - 2$, we get $b \le r - 2 = a$. If $a \ge r-1$, then $b \le a$ is also satisfied since $b \le r - 1$. Thus

$$\left| \left\{ (u,v) \;\middle|\; 0 \le v \le \min(u,r-1)\,,\, 0 \le ur - v \le 2g-1 \right\} \right| = \alpha - 2g + 1. \tag{2}$$

Now, consider the first set :
Let $0 \le k \le 2g-1$ and $a, b \in \mathbf{N}$ such that $k = ar - b$. Since $2g - 1 = r(r-2) - (r-1)$, this number cannot be represented by $ar - b$, where (a,b) belongs to the first set. So, let $0 \le k \le 2g-2$. This implies that $k < r(r-2) - (r-1)$, hence $a \le r - 3 < r - 1$. Thus, for fixed $0 \le a \le r-3$, we get $a + 1$ elements (a,b) of the first set, as b runs over the values $0, \ldots, a$. Hence

$$\left| \left\{ (u,v) \;\middle|\; 0 \le v \le \min(u,r-1)\,,\, 2g \le ur - v \le \alpha \right\} \right| = \sum_{i=1}^{r-2} i = \frac{1}{2}(r-2)(r-1) = g. \tag{3}$$

Adding (2) and (3), we get

$$\left| \left\{ (u,v) \;\middle|\; 0 \le v \le \min(u,r-1)\,,\, 0 \le ur - v \le \alpha \right\} \right| = \alpha - g + 1.$$

On the other hand, by the Riemann-Roch theorem, we have

$$l(\alpha Q) = \alpha - g + 1,$$

which proves (*), and hence, the assertion of theorem 1.

Q.E.D.

Remark :

Consider the set $M_\alpha := \{(u,v) \mid 0 \le ur - v \le \alpha\,,\, 0 \le v \le u\}$, and define on M_α the equivalence relation $\sim$ by

$$(u,v) \sim (u',v') \overset{\text{def}}{\iff} ur - v = u'r - v'.$$

Let $\mathcal{M}_\alpha$ denote a complete set of representatives of $M_\alpha / \sim$. Then, one can prove in the same way as above :

Theorem 2'. *The set*

$$\left\{ \frac{y^v z^{u-v}}{(x - \varsigma z)^u} \;\middle|\; (u,v) \in \mathcal{M}_\alpha \right\}$$

is a basis of $L(\alpha Q)$.

This form of theorem 2 is sometimes useful if one wants to construct a basis of $L(\alpha Q)$ which contains many conjugate elements (cf. section 4).

4. Subcodes over the prime field

Let $h \in \mathbf{N}$, $q = p^h$ a prime power, and $\mathbf{V} \leq GF(q)^n$ a vector space of dimension k. Denote by σ the Frobenius automorphism of $GF(q)/GF(p)$, and by $\mathcal{G}$ the Galois group $Gal\big(GF(q)/GF(p)\big)$, i.e. $\mathcal{G} = \langle\sigma\rangle$. $\mathcal{G}$ acts on the set $GF(q)^n$ by acting on the coordinates of the vectors in $GF(q)^n$:

$$\underline{x}^\tau = (x_1, \dots, x_n)^\tau := (x_1{}^\tau, \dots, x_n{}^\tau), \qquad x_i \in GF(q)\,, \tau \in \mathcal{G}.$$

Theorem 3. *Let* $\mathbf{V}$ *be invariant under the action of* $\mathcal{G}$. *Then* $\mathbf{V}$ *possesses a basis* $\{\underline{b}_1, \dots, \underline{b}_k\}$ *where* $\underline{b}_i \in GF(p)^n$ *for* $i = 1, \dots, k$.

Proof. Let $\{\underline{a}_1, \dots, \underline{a}_k\}$ be an arbitrary basis of $\mathbf{V}$. Further, choose $\beta \in GF(q)$ such that $\{1, \beta, \dots, \beta^{h-1}\}$ is a basis of $GF(q)/GF(p)$ (for example, by choosing β as a zero of the prime polynomial which generates $GF(q)$). For $\underline{a} \in \mathbf{V}$, let $\underline{a}^{(i)}$ denote $\underline{a}^{\sigma^i}$ and consider the span $\langle \underline{a}_1, \underline{a}_1{}^{(1)}, \dots, \underline{a}_1{}^{(h-1)}, \dots, \underline{a}_k, \dots, \underline{a}_k{}^{(h-1)} \rangle$. Define $\underline{\tilde{b}}_{ij} \in \mathbf{V}$ by

$$\underline{\tilde{b}}_{ij} := \sum_{\nu=1}^{h-1} \left(\beta^{j-1} \cdot \underline{a}_i\right)^{(\nu)}, \qquad 1 \leq i \leq k\,, 1 \leq j \leq h.$$

Clearly, $\underline{\tilde{b}}_{ij} \in GF(p)^n$ since their coordinates are traces of elements in $GF(q)$. Furthermore, since the Vandermonde matrix

$$\begin{pmatrix} 1 & 1 & \dots & 1 \\ \beta & \beta^{(1)} & \dots & \beta^{(h-1)} \\ \vdots & \vdots & \ddots & \vdots \\ \beta^{h-1} & \left(\beta^{(1)}\right)^{h-1} & \dots & \left(\beta^{(h-1)}\right)^{h-1} \end{pmatrix}$$

is not singular, we have for all $i = 1, \dots, k$: $\underline{a}_i \in \langle \underline{\tilde{b}}_{i1}, \dots, \underline{\tilde{b}}_{ih} \rangle$; thus

$$\mathbf{V} = \left\langle \underline{\tilde{b}}_{ij} \;\middle|\; 1 \leq i \leq k\,, 1 \leq j \leq h \right\rangle.$$

Q.E.D.

Remark :

One can actually prove by the same method as above :

Theorem 3'. *Let* F *be a finite field,* K/F *a Galois extension of* F *with the Galois group* $\mathcal{G}$. *Further, let* $\mathbf{V} \leq K^n$ *be a vector space over* K *which is invariant under the action of* $\mathcal{G}$. *Then* $\mathbf{V}$ *posseses an* F*-rational basis.*

One can use theorem 3 in order to construct subcodes over the prime field in the following way :

Let C be a code defined over $GF(q)$ and let $C^{\perp}$ denote its dual code generated by $\underline{a}_1, \ldots, \underline{a}_r$. Then

$$(\tilde{C^{\perp}}) := \langle \underline{a}_1, \ldots, \underline{a}_1^{(h-1)}, \ldots, \underline{a}_r, \ldots, \underline{a}_r^{(h-1)} \rangle$$

is the vector space of least dimension containing $C^{\perp}$ and being invariant under the action of $\mathcal{G}$. Let $\tilde{C}$ be defined by $\tilde{C}^{\perp} := (\tilde{C^{\perp}})$. Then $\tilde{C} \leq C$. One can find a $GF(p)$-rational basis of $\tilde{C}$ by theorem 3. Consider the $GF(p)$-span of this basis; this is a code over $GF(p)$, say C'. Clearly, $C' \leq \tilde{C} \leq C$ and hence, the minimal distance $d(C')$ of C' satisfies

$$d(C') \geq d(C). \tag{4}$$

If C is an $\big(n, k, d(C)\big)$-code, one gets

$$\dim C' \geq hk - n(h-1). \tag{5}$$

5. Asymptotics of subcodes over the prime field

Let $h \in \mathbf{N}$ and $\mathbf{X}_h$ denote the Hermitian curve $x^{p^h+1} + y^{p^h+1} + z^{p^h+1} = 0$ over the field $GF(p^{2h})$. Choose a sequence $\{\alpha_h\}_{h \in \mathbf{N}}$ such that $2g(\mathbf{X}_h) - 2 < \alpha_h < \overline{n}(\mathbf{X}_h) - 1$, and denote by C_h the Hermitian Goppa code corresponding to the space $\Omega(\alpha_h Q_h - \sum P_i^{(h)})$ where Q_h is a $GF(p)$-rational point of $\mathbf{X}_h$ and the $P_i^{(h)}$'s are the other $GF(p^{2h})$-rational points of $\mathbf{X}_h$. Let $\tilde{C}_h$ denote the subcode of C_h over the prime field; the redundancy $\tilde{r}_h$ of $\tilde{C}_h$ satisfies by (5) :

$$\tilde{r}_h \leq \min(2hr_h, n_h)$$

where r_h denotes the redundancy of C_h. Define $\psi : \{\alpha_h | h \in \mathbf{N}\} \to \mathbf{Q}$ by $\alpha_h \mapsto \psi(\alpha_h) := \tilde{r}_h / r_h$. ψ is well defined because, by (1), r_h is a linear function of α_h. Furthermore, one has

$$\forall\, h \in \mathbf{N} \qquad 1 \leq \psi(\alpha_h) \leq \min\left(2h, \frac{n_h}{r_h}\right). \tag{6}$$

For the dimension $\tilde{k}_h$ of $\tilde{C}_h$ one gets by (5) :

$$\tilde{k}_h = \big(\psi(\alpha_h) - 1\big)\big(k_h - n_h\big) + k_h, \tag{7}$$

where $n_h := \overline{n}(\mathbf{X}_h) - 1$ is the length of the codewords of C_h, and $g_h := g(\mathbf{X}_h)$. (4) and (1) give the following estimation for the minimal distance $\tilde{d}_h$ of $\tilde{C}_h$ in terms of the minimal distance d_h of C_h

$$\tilde{d}_h \geq d_h \geq \alpha_h - 2g_h + 2. \tag{8}$$

Thus, (1) implies :

$$\tilde{d}_h + \tilde{k}_h \geq \big(\psi(\alpha_h) - 1\big)\big(-\alpha_h + g_h - 1\big) + n_h - g_h + 1. \tag{9}$$

Let $\tilde{R}_h$ and $\tilde{\delta}_h$ denote the values $\tilde{k}_h/n_h$ and $\tilde{d}_h/n_h$ respectively. We are interested is the following question: What happens with $\tilde{R}_h$ and $\tilde{\delta}_h$, as $h \to \infty$?
Inequality (9) implies :

$$\lim_{h\to\infty} (\tilde{\delta}_h + \tilde{R}_h) =: \tilde{\delta} + \tilde{R} \geq 1 - \lim_{h\to\infty} \left((\psi(\alpha_h) - 1) \cdot \frac{\alpha_h}{n_h} \right).$$

Set $\gamma := \lim_{h\to\infty} (\psi(\alpha_h) - 1)\alpha_h/n_h$. Then the line $\tilde{\delta} + \tilde{R} = 1 - \gamma$ intersects the Gilbert-Varshamov curve iff $\gamma < \log_p(2p-1) - 1$. Let us denote by $\tilde{\delta}_1(\gamma)$,$\tilde{\delta}_2(\gamma)$ the $\tilde{\delta}$-values of the intersection, and assume that $\tilde{\delta}_1(\gamma) < \tilde{\delta}_2(\gamma)$. By (8), one has :

$$\tilde{\delta} \geq \lim_{h\to\infty} \frac{\alpha_h}{n_h}.$$

In other words : If one can find a sequence $\{\alpha_h\}_{h\in N}$ with

$$\gamma := \lim_{h\to\infty} (\psi(\alpha_h) - 1)\frac{\alpha_h}{n_h} < \log_p(2p-1) - 1,$$

and $\lim_{h\to\infty} \alpha_h/n_h \geq \tilde{\delta}_1(\gamma)$, then, this sequence induces a sequence of codes which are asymptotically better than the Gilbert - Varshamov bound.

Acknowledgement:

The author is indebted to *Prof. H. W. Leopoldt* and *Prof. Th. Beth* for their very helpful comments and hints. The author also wants to thank *Dr. M. Clausen* for the many useful discussions which led to the present form of this paper.

References :

1. V. D. GOPPA, *Codes and Information*, Russian Math. Survey 39,1 (1984), 87-141.
2. H. HASSE, *Number Theory*, Springer-Verlag, 1980.

SYMMETRIES OF CYCLIC EXTENDED GOPPA CODES OVER $\mathbb{F}_q$

J.A. THIONG-LY
University of Toulouse le Mirail
5, av. Antonio Machado - 31058 Toulouse
France

Abstract : We use elements in the group $PGL(2,q^m)$ to define the location sets of cyclic extended Goppa Codes over $\mathbb{F}_q$, where the lengths of the codes divide $q^m \pm 1$.

0 - INTRODUCTION

Let $L' = \{\alpha_1,\ldots,\alpha_n\}$ be a subset of the Galois field $\mathbb{F}_{q^m}$ and let $g(z)$ be a polynomial over $\mathbb{F}_{q^m}$ with roots not in L' and having degree less than n.

A Goppa Code ([1], [6]) is the set of n-dimensional vectors $(X_{\alpha_1},\ldots,X_{\alpha_n})$ over $\mathbb{F}_q$ which satisfy :

$$\sum_{\alpha_i \in L'} \frac{X_{\alpha_i}}{z - \alpha_i} \equiv 0 \text{ modulo } g(z) .$$

Such a code is denoted by $\Gamma(L',g)$. L' is called the location set and g the Goppa polynomial. The extended code of code $\Gamma(L',g)$ is the set of (n+1)-tuples $(X_\infty,X_{\alpha_1},\ldots,X_{\alpha_n})$ such that :

$$(X_{\alpha_1},\ldots,X_{\alpha_n}) \in \Gamma(L',g) \text{ and } X_\infty = - \sum_{\alpha_i \in L'} X_{\alpha_i} .$$

We denoted the extended Goppa code by $\Gamma(L,g)$ where $L = L' \cup \{\infty\}$. In the following we consider the extended Goppa codes :

1) $\Gamma_1 = \Gamma(L_1,g_1)$ where :

$$g_1 = (z - \beta_1)^a(z - \beta_2)^a,$$

here β_1 and β_2 are two conjugate elements in $\mathbb{F}_{q^{2m}} \setminus \mathbb{F}_{q^m}$, and a is an integer ≥ 1.

L_1 is some subset of $\mathbb{F}_{q^m} \cup \{\infty\}$, containing ∞, and such that $|L_1|$ divides $q^m + 1$.

2) $\Gamma_2 = \Gamma(L_2,g_2)$ where :

$$g_2 = (z - \beta_1)^{a_1}(z - \beta_2)^{a_2},$$

here β_1 and β_2 are two distinct elements in $\mathbb{F}_{q^m}$ and a_1, a_2 are two integers ≥ 1.

L_2 is some subset in $(\mathbb{F}_{q^m} \cup \{\infty\}) \setminus \{\beta_1,\beta_2\}$, containing ∞, and such that $|L_2|$ divides $q^m - 1$.

3) $\Gamma_3 = \Gamma(L_3,g_3)$ where

$$g_3 = (z - \beta)^a ,$$

here β is an element in $\mathbb{F}_{q^m}$ and a is an integer ≥ 1.

L_3 is some subset of $\mathbb{F}_{q^m} \cup \{\infty\}$, containing ∞, and such that $|L_3|$ divides $q^m - 1$.

In case 1, for $q = 2$, $a = 1$, $L_1 = \mathbb{F}_{2^m} \cup \{\infty\}$, Berlekamp and Moreno ([0]) proved that these codes are cyclic codes. Their proofs uses symmetries of the code.

Codes Γ_1 with $|L_1| = q^m+1$ and codes Γ_2 with $|L_2| = q^m-1$ are studied by Tzeng and Zimmerman in [3]. They prove that these codes are cyclic and reversible, that is if

$(X_0, X_A, \ldots, X_{N-1})$ is a codeword, then

$(X_{N-1}, \ldots, X_1, X_0)$ is also a codeword.

All their proofs are based on a special form of the parity check matrix. Now we want to make some remarks on the authors references [4] and [5].

1) The Tzeng and Yu proof of theorem 1 (see [4]) make use of a parity check matrix depending of the Goppa polynomial. Note that this parity check matrix is defined over a large field and that the code which is considered actually is the subfield subcode. At a certain point in the proof it is remarked that the given parity check matrix is of the form :

$$\begin{bmatrix} 1 & 1 & 1 & \cdots & 1 \\ 1 & \gamma_1 & \gamma_1^2 & \cdots & \gamma_1^{N-1} \\ 1 & \gamma_2 & \gamma_2^2 & \cdots & \gamma_2^{N-1} \\ \vdots & \vdots & \vdots & \vdots & \vdots \end{bmatrix}$$

Can we conclude that two parity check matrices are equivalent if we only know that the corresponding subfield subcodes are equivalent ? According to we this is not the case and there are a lot of counterexamples.

2) The Vishnevetskii proof of his theorem 7 (see [5]) states : "since the code Γ is invariant relative to a cyclical permutation of coordinates, the coordinates can be rearranged in such a way that the resultant code is Θ-invariant".

According to condition A of [5], Θ is some fixed element in $G\backslash\{1\}$ such that L is a Θ-orbit.

Now why can the coordinates be rearranged in such a way that the resultant code is Θ-invariant ? Note that we does not know the ordering on the location set by which the code is a cyclic code. The author seems to use the following ordering : $\infty \to \Theta\,\infty \to \ldots\ \Theta^{r-1}\infty$. So, it appears that the following result (cf. corollary 12 of [5]) needs a proof :

Theorem 1 : an extended Goppa Code $\Gamma(L,g)$ where $L = \mathbb{F}_{q^m} \cup \{\infty\}$ is a cyclic code if and only if there exists $g' = (z-\beta_1)^a(z-\beta_2)^a$ where β_1 and β_2 are conjugate elements in $\mathbb{F}_{q^{2m}} \backslash \mathbb{F}_{q^m}$ and a an integer ≥ 1 such that $\Gamma(L,g) = \Gamma(L,g')$.

A proof of this theorem will be given. For code over $\mathbb{F}_{q^m}$, we give in theorem 4 an improvement of this theorem.

I - Some properties about actions of the group $G = PGL(2,q^m)$ ([5], [6])

Let $G = PGL(2,q^m)$. Then G acts on $\mathbb{F}_{q^m} \cup \{\infty\}$ as follows :

if $p = \begin{pmatrix} a & b \\ c & d \end{pmatrix} \in G$ then $px = \frac{ax+b}{cx+d}$, $(x \in \mathbb{F}_{q^m} \cup \{\infty\})$.

The set $\mathbb{F}_{q^m} \cup \{\infty\}$ can be viewed as a projective line : ∞ corresponds to the point $\begin{pmatrix} u \\ 0 \end{pmatrix}$; $x \in \mathbb{F}_{q^m}$ corresponds to the point $\begin{pmatrix} x \\ 1 \end{pmatrix}$.

The above described action of G on $\mathbb{F}_{q^m} \cup \{\infty\}$ is the same as the action on this projective line by matrix multiplication :

$$\begin{pmatrix} a & b \\ c & d \end{pmatrix}\begin{pmatrix} x \\ 1 \end{pmatrix} = \begin{pmatrix} ax+b \\ cx+d \end{pmatrix}.$$

It is well known that G acts 2-transitive on the set $\mathbb{F}_{q^m}$. The action of G on $\mathbb{F}_{q^m} \cup \{\infty\}$ can be extended to $\overline{\mathbb{F}}_{q^m} \cup \{\infty\}$ where $\overline{\mathbb{F}}_{q^m}$ is the algebraic closure of $\mathbb{F}_{q^m}$.

Moreover, G acts on $\mathbb{F}_{q^m}[z]$ in the following way :

if $p = \begin{pmatrix} a & b \\ c & d \end{pmatrix} \in G$ and $f(z) \in \mathbb{F}_{q^m}[z]$, then

$$pf(z) = (cz+d)^r \, f(\frac{az+b}{cz+d}), \text{ where } r \text{ is the degree of } f.$$

Polynomials f and h are said to be equivalent (denoted : $f \sim h$) if $f = \lambda h$ for some $\lambda \in \mathbb{F}^*_{q^m}$.

Note that $pfh = pf\ ph$ (f and h in $\mathbb{F}_{q^m}[z]$).

Therefore this action of p does not affect the multiplicities of the roots of a polynomial.

Property 1 : Let $f \in \mathbb{F}_{q^m}[z]$ and $p \in G$.

$pf \sim f$ if and only if $pR = R$ where R is the set of roots of f in $\overline{\mathbb{F}}_{q^m}$.

Now an element $p \in G$ is a symmetrie of the code $\Gamma(L,g)$ if for each codeword $(x_{\alpha_i})_{\alpha_i \in L}$, also $(x_{p\alpha_i})_{\alpha_i \in L}$ is a codeword.

Property 2 : Let $p \in G$ and $pL = L$.

p is a symmetrie of the code $\Gamma(L,g)$ if and only if $pg \sim g$.

Proofs of these properties may be found in [5].

Now : given Goppa polynomials g_1, g_2, g_3 (of the form 1), 2), 3)).
given $|L_1|$, $|L_2|$, $|L_3|$ such that $|L_1|$ divides q^m+1, $|L_2|$ and $|L_3|$ divide q^m-1, we can construct location sets L_1, L_2, L_3 such that the codes Γ_1, Γ_2 and Γ_3 are cyclic codes, (not merely equivalent to cyclic codes). For this, it is sufficient to exhibit a symmetrie π of order $r_i = |L_i|$ and take for L_i :

$$L_i = \{\infty, \pi\infty, \ldots, \pi^{r_i-1}\infty\} \qquad \text{(with this ordering).}$$

Moreover, for codes Γ_1 and Γ_2, we determine a group of symmetries which contains an involution transforming each codeword into the reversed codeword.

Notice that extended codes whose length is a proper divisor of q^m+1 were not investigated by the previous authors.

II - A group of symmetries for the codes Γ_1, Γ_2 and Γ_3

Given the elements β_1 and β_2, let $\Theta(\eta)$, $\pi(\eta)$ and ε be the following matrices :

$$\Theta(\eta) = \begin{pmatrix} \eta & b \\ 1 & -\eta \end{pmatrix} \qquad \text{where } b = \beta_1\beta_2 - \eta(\beta_1+\beta_2),$$

$$\pi(\eta) = \begin{pmatrix} \beta_1+\beta_2-\eta & -\beta_1\beta_2 \\ 1 & -\eta \end{pmatrix} ,$$

$$\varepsilon = \begin{pmatrix} -1 & \beta_1+\beta_2 \\ 0 & 1 \end{pmatrix} .$$

Lemma 1 :

1) For any η we have : $\pi(\eta) = \varepsilon.\Theta(\eta)$.
2) For any η, η' we have : $\pi(\eta)\Theta(\eta')\pi(\eta) = (b+\eta^2)\Theta(\eta')$.
3) Let β_1 and β_2 two distinct elements in $\mathbb{F}_{q^m}$. Then for any $\eta \in \mathbb{F}_{q^m}\setminus\{\beta_1,\beta_2\}$, the matrices $\Theta(\eta)$ and $\pi(\eta)$ belong to $PGL(2,q^m)$, and $\Theta^2 = (b+\eta^2)Id$.
4) Let β_1 and β_2 two conjugate elements in $\mathbb{F}_{q^{2m}}\setminus\mathbb{F}_{q^m}$. Then for any $\eta \in \mathbb{F}_{q^m}$ the matrices $\Theta(\eta)$ and $\pi(\eta)$ belong to $PGL(2,q^m)$ and $\Theta^2 = (b+\eta^2)Id$.

Proofs : direct.

Lemma 2 :

Let β_1 and β_2 two distinct elements in $\mathbb{F}_{q^m}$ or in $\mathbb{F}_{q^{2m}}\setminus\mathbb{F}_{q^m}$.

Let $\pi' = \begin{pmatrix} \gamma_1+\gamma_2 & -\gamma_1\gamma_2 \\ 1 & 0 \end{pmatrix}$.

Then the order of π' is equal to the order of $\frac{\gamma_1}{\gamma_2}$.

Proof . The characteristic polynomial of π' is $(\pi-\gamma_1)(\pi-\gamma_2)$. Therefore

$$\pi' = \upsilon \begin{pmatrix} \gamma_1 & 0 \\ 0 & \gamma_2 \end{pmatrix} \upsilon^{-1} \quad \text{for some } \upsilon.$$

When γ_1 and γ_2 are elements in $\mathbb{F}_{q^m}$ (respectively in $\mathbb{F}_{q^{2m}} \setminus \mathbb{F}_{q^m}$),

$$\begin{pmatrix} \gamma_1 & 0 \\ 0 & \gamma_2 \end{pmatrix} \text{ and } \begin{pmatrix} \frac{\gamma_1}{\gamma_2} & 0 \\ 0 & 1 \end{pmatrix} \quad \text{represent the same element in } PGL(2,q^m)$$

(respectively in $PGL(2,q^{2m})$).

Therefore the order of π' is equal to the order of $\frac{\gamma_1}{\gamma_2}$ and divides q^m-1 or q^m+1 depending on γ_1 and γ_2 belonging to $\mathbb{F}_{q^m}$ or $\mathbb{F}_{q^{2m}}$ □ .

Our construction of location sets L_i is based on the following lemma :

Lemma 3 : Let $\pi_\gamma = \pi\left(\frac{\beta_1-\gamma\beta_2}{1-\gamma}\right)$:

$$\pi_\gamma = \begin{pmatrix} \beta_1+\beta_2 - \frac{\beta_1-\gamma\beta_2}{1-\gamma} & -\beta_1\beta_2 \\ 1 & -\frac{\beta_1-\gamma\beta_2}{1-\gamma} \end{pmatrix}$$

1) If β_1 and β_2 are two conjugate elements in $\mathbb{F}_{q^{2m}} \setminus \mathbb{F}_{q^m}$ and if $\gamma \in \mathbb{F}_{q^{2m}} \setminus \mathbb{F}_{q^m}$ is a rth root of unity in $\mathbb{F}_{q^{2m}}$ such that $r | q^m+1$, then π_γ belongs to $PGL(2,q^m)$ and has the same order as γ.

2) If β_1 and β_2 are two distinct elements in $\mathbb{F}_{q^m}$ and $\gamma \in \mathbb{F}_{q^m}$ is a rth root of unity in $\mathbb{F}_{q^m}$, then π_γ belongs to $PGL(2,q^m)$ and has the same order as γ.

Proof : Let $\gamma = \dfrac{\gamma_1}{\gamma_2}$.

If we choose

$$\Theta = \begin{pmatrix} \dfrac{\gamma_1-\gamma_2}{\beta_1-\beta_2} & \dfrac{\gamma_2\beta_1-\gamma_1\beta_2}{\beta_1-\beta_2} \\ 0 & 1 \end{pmatrix}$$

then $\Theta\beta_i = \gamma_i \qquad (i = 1,2)$.

Let $$\pi' = \begin{pmatrix} \gamma_1+\gamma_2 & -\gamma_1\gamma_2 \\ 1 & 0 \end{pmatrix} .$$

Since $\pi'\gamma_i = \gamma_i \quad (i = 1,2)$, we have $\Theta^{-1}\pi'\Theta\beta_i = \beta_i \quad (i = 1,2)$.

Let $\pi = \Theta^{-1}\pi'\Theta$.

By lemma 2, the order of π is equal to the order of γ.
By direct calculations we have :

$$\pi = \begin{pmatrix} \beta_1+\beta_2 + \dfrac{\beta_1\gamma_2-\beta_2\gamma_1}{\gamma_1-\gamma_2} & -\beta_1\beta_2 \\ 1 & \dfrac{\beta_1\gamma_2-\beta_2\gamma_1}{\gamma_1-\gamma_2} \end{pmatrix} = \pi_\gamma .$$

So $\pi_\gamma = \pi(\eta_0)$ with $\eta_0 = \dfrac{\beta_1-\gamma\beta_2}{1-\gamma}$.

Case 2. By construction, the coefficients of π_γ are in $\mathbb{F}_{q^m}$. It is easy to verify that $\eta_0 \notin \{\beta_1,\beta_2\}$. Therefore by lemma 1, $\pi_\gamma \in PGL(2,q^m)$.

Case 1. Since β_1 and β_2 are conjugate elements, $\beta_1+\beta_2$ and $\beta_1\beta_2$ are elements in $\mathbb{F}_{q^m}$.

Now :

$$\left(\frac{\beta_1-\gamma\beta_2}{1-\gamma}\right)^{q^m} = \frac{\beta_2-\beta_1\gamma^{q^m}}{1-\gamma^{q^m}} = \frac{\gamma\beta_2-\beta_1\gamma^{q^m+1}}{\gamma-\gamma^{q^m+1}}$$

$$= \frac{\beta_1-\gamma\beta_2}{1-\gamma} \quad \text{since } \gamma^{q^m+1} = 1 \quad .$$

So $$\frac{\beta_1-\gamma\beta_2}{1-\gamma} \in \mathbb{F}_{q^m} .$$

therefore $\pi_\gamma \in PGL(2,q^m)$. □ .

Since $\pi_\gamma\beta_i = \beta_i$ $(i = 1,2)$, from property 2 and lemma 3, we have directly the following corollaries :

Corollary 1 : Choose as Goppa polynomial $g_1(z) = (z-\beta_1)^a(z-\beta_2)^a$ where β_1 and β_2 are two conjugate elements in $\mathbb{F}_{q^{2m}} \setminus \mathbb{F}_{q^m}$, and a an integer ≥ 1.

Let $r = \frac{q^m+1}{s}$, and let γ be a primitive r^{th} root of unity in $\mathbb{F}_{q^{2m}} \setminus \mathbb{F}_{q^m}$.

Then $\Gamma(L,g_1)$ with

$$L_1 = \{\infty, \pi_\gamma\infty, \ldots, \pi_\gamma^{r-1}\infty\}$$

is an extended cyclic Goppa code with length r.

Example 1 : (see also [6] page 342).

Let : $L_1 = \mathbb{F}_{2^3} \cup \{\infty\}$, $\mathbb{F}_{2^6} = \mathbb{F}_{2^3}(\mu)$, $\mathbb{F}_{2^3} = \mathbb{F}_2(\alpha)$.

Choose $g_1 = (z-\mu^{21})(z-\mu^{42}) = z^2+z+1$, $\gamma = \mu^7$.

The minimum polynomial of μ is : $X^2 + \alpha X + \alpha^5$.

One verifies that :

$$\pi_\gamma = \begin{pmatrix} \alpha^2 & 1 \\ 1 & \alpha^6 \end{pmatrix}.$$

The following ordering of $\mathbb{F}_{2^3} \cup \{\infty\}$ can be choose in order $\Gamma(L_1,g_1)$ be cyclic :

$$\infty \to \alpha^2 \to \alpha^5 \to 0 \to \alpha \to \alpha^3 \to 1 \to \alpha^4 \to \alpha^6$$

The codewords are :

0	0	0	0	0	0	0	0	0
0	1	1	0	1	1	0	1	1
1	0	1	1	0	1	1	0	1
1	1	0	1	1	0	1	1	0

One verifies that the code is also a reversible code. The reason of this will be given later.

Corollary 2 : Choose as Goppa polynomial $g(z) = (z-\beta_1)^{a_1}(z-\beta_2)^{a_2}$ where β_1 and β_2 are two distinct elements in $\mathbb{F}_{q^m}$ and a_1, a_2 two integers ≥ 1.

Let $r = \frac{q^m-1}{s}$ and let γ be a primitive r^{th} root of unity in $\mathbb{F}_{q^m}$. Then $\Gamma(L_2,g_2)$ with

$$L_2 = \{\infty, \pi_\gamma \infty, \ldots, \pi_\gamma^{r-1}\infty\}$$

is a cyclic code.

Denote by $\pi_{\gamma,x}$ the matrix π_γ of lemma 3 where β_1 is substituded by β and where β_2 is substituded by any element x not equal to β.

Corollary 3 : Choose as Goppa polynomial $g_3(z) = (z-\beta)^a$ where β is an element in $\mathbb{F}_{q^m}$ and a an integer ≥ 1.

Let $r = \frac{q^m-1}{s}$ and let γ be a primitive r^{th} root of unity in $\mathbb{F}_{q^m}$. Then $\Gamma(L_3, g_3)$ with

$$L_3 = \{\infty\ ,\ \pi_{\gamma,x}\infty, \ldots\}$$

is a cyclic code.

Example 2. Choose $g_3(z) = (z-\alpha^3) \in \mathbb{F}_{3^2}[z]$.

Let $r = 4$ and $\gamma = \alpha^2$. For $x = \alpha^6$ we have $\pi = \begin{pmatrix} \alpha^2 & \alpha^5 \\ 1 & \alpha^5 \end{pmatrix}$. The code $\Gamma(L_3, g_3)$ with the location set $L = \{\infty\ ,\alpha^2,\alpha^4,\alpha\}$ is a cyclic code.

The codewords are : 0000, 2121, 1212 .

Remark : The sets L_x defined by Tzeng and Yu in [4] are merely

$L_x = \{p\infty, \ldots, p^{r-1}\infty\}$ where $p = \Pi_{\gamma, x-\beta}$.

Theorem 1 :

An extended Goppa code $\Gamma(L,g)$ where $L = \mathbb{F}_{q^m} \cup \{\infty\}$ is a cyclic code if and only if there exists $g' = (z-\beta_1)^a(z-\beta_2)^a$ where β_1 and β_2 are conjugate elements in $\mathbb{F}_{q^{2m}} \setminus \mathbb{F}_{q^m}$ and a an integer ≥ 1, such that $\Gamma(L,g) = \Gamma(L,g')$.

Proof : Suppose $\Gamma(L,g)$ be a cyclic code with location set $L = \mathbb{F}_{q^m} \cup \{\infty\}$, and let γ be a primitive N^{th} root of unity in $\mathbb{F}_{q^{2m}}$ $(N = q^m+1)$.
Now, consider the code $\mathbf{C} = \Gamma(L,h = (z-\beta_1)(z-\beta_2))$ where β_1 and β_2 are two conjugate elements in $\mathbb{F}_{q^{2m}}$.

$$\text{Let} \qquad \pi_\gamma = \begin{pmatrix} \beta_1+\beta_2 - \dfrac{\beta_1-\gamma\beta_2}{1-\gamma} & -\beta_1\beta_2 \\ 1 & -\dfrac{\beta_1-\gamma\beta_2}{1-\gamma} \end{pmatrix} \in G \ .$$

By corollary 1, $\mathbf{C}$ is a cyclic code with the location set

$$L = \{\infty, \pi_\gamma\infty, \ldots, \pi_\gamma^{N-1}\infty\} = \mathbb{F}_{q^m} \cup \{\infty\} .$$

Denote M_γ the matrix $\begin{pmatrix} \gamma & 0 \\ 0 & 1 \end{pmatrix}$: M_γ represents the multiplication by γ.

It is easy to verify that : $M_\gamma = \varphi . \pi_\gamma . \varphi^{-1}$ where $\varphi = \begin{pmatrix} -1 & \beta_2 \\ -1 & \beta_1 \end{pmatrix}$.

Let $C = (C_1, C_\gamma, \ldots, C_{\gamma^{N-1}})$ be a codeword in the cyclic code $\Gamma(L,g)$.

Now, we may number the coordinates by the set $\{\infty, \pi\infty, \ldots, \pi^{N-1}\infty\}$ and write any codework C in $\Gamma(L,g)$

$$C = (C_\infty, C_{\pi\infty}, \ldots, C_{\pi^{N-1}\infty}) .$$

But, by construction of $\Gamma(L,h)$, C verifies :

$$\frac{C_{\pi_\infty}}{z-\pi_\infty} + \ldots + \frac{C_{\pi^{N-1}_\infty}}{z-\pi^{N-1}_\infty} = 0 \bmod h$$

and $C_\infty = -(C_{\pi\infty} + \ldots + C_{\pi^{N-1}\infty})$.

Therefore, $\Gamma(L,g)$ is included in $\Gamma(L,h)$, and $\Gamma(L,g) = \Gamma(L,g')$ where $\Gamma(L,g')$ is a cyclic Goppa code included in $\Gamma(L,h)$. It is easy to prove that g' is necessarly of the form $g' = h^a$ for some integer $a \geq 1$. Converse of the proof follows directly by the corollary 1. □ .

Now, let $\Theta_\gamma = \Theta(\eta)$ with $\eta = \dfrac{\beta_1-\gamma\beta_2}{1-\gamma}$:

$$\Theta_\gamma = \begin{pmatrix} \dfrac{\beta_1-\gamma\beta_2}{1-\gamma} & b \\ 1 & -\dfrac{\beta_1-\gamma\beta_2}{1-\gamma} \end{pmatrix} \qquad b = \beta_1\beta_2 - (\beta_1+\beta_2)\eta.$$

Let : $\quad \varepsilon = \begin{pmatrix} -1 & \beta_1+\beta_2 \\ 0 & 1 \end{pmatrix}$ and

$$L_i = \{\infty, \pi_\gamma\infty, \ldots, \pi_\gamma^{r-1}\infty\}$$

the location set for the code Γ_i $(i = 1,2)$.

With the same hypothesis as in corollaries 1 and 2 for elements β_1, β_2 and γ, we have the following theorem :

Theorem 2 : **1)** Codes Γ_1 and Γ_2 are reversible codes.

2) The group generated by ε and by π_γ is a group of symmetries of order $2r$ for the codes Γ_1 and Γ_2.

Proof : **1)** By the same arguments given in lemma 3 for π_γ, the coefficients of Θ_γ belong to $\mathbb{F}_{q^m}$.

Since $\dfrac{\beta_1-\gamma\beta_2}{1-\gamma} \notin \{\beta_1,\beta_2\}$ (for else $\beta_1 = \beta_2$), by lemma 1, Θ_γ is an element of $PGL(2,q^m)$.

Let $\quad L_i = \{\infty, \pi_\gamma\infty, \ldots, \pi_\gamma^{r-1}\infty\}$.

One verifies that : $\Theta_\gamma\infty = \pi_\gamma^{-1}\infty$.

But by lemma 1-2), we have : $\Theta_\gamma\pi_\gamma = \pi_\gamma^{-1}\Theta_\gamma$.

So, for $i = 0, \ldots, r-1$ we have : $\Theta_\gamma\pi_\gamma^{i}\infty = \pi_\gamma^{-i-1}\infty$.

This proves that Θ_γ is a reversible symmetry for codes Γ_1 and Γ_2 (recall that $\Theta_\gamma\beta_1 = \beta_2$ and $\Theta_\gamma\beta_2 = \beta_1$).

2) Let W be the group generated by Θ_γ and by π_γ. Since $\Theta_\gamma^2 = \mathrm{Id}$,

$\pi_\gamma^r = \mathrm{Id}$ and $\Theta_\gamma\pi_\gamma = \pi_\gamma^{-1}\Theta_\gamma$, we have : $W = \{\Theta_\gamma^i \, \pi_\gamma^j \mid i = 0,1 \text{ and } j = 0,\dots,r-1\}$

and $|W| = 2r$.

Besides, one verifies that $\Theta_\gamma\pi_\gamma = \begin{pmatrix} -1 & \beta_1+\beta_2 \\ 0 & 1 \end{pmatrix} = \varepsilon \in W$.

Therefore $W = \{\varepsilon^i \pi_\gamma^j \mid i = 0,1 \text{ and } j = 0,\dots,r-1\}$. □

Example 3 :

For the code of example 1 ($\beta_1 = \mu^{21}$, $\beta_2 = \mu^{42}$, $\gamma = \mu^7$) we find

$$\Theta_\gamma = \begin{pmatrix} \alpha^6 & \alpha^2 \\ 1 & \alpha^6 \end{pmatrix}.$$

One verifies that Θ_γ reverses the ordering of the elements of L :

$$\begin{array}{ccccccccc} \infty & \alpha^2 & \alpha^5 & 0 & \alpha & \alpha^3 & 1 & \alpha^4 & \alpha^6 \\ \Theta_\gamma \downarrow & & & & & & & & \downarrow \\ \alpha^6 & \alpha^4 & 1 & \alpha^3 & \alpha & 0 & \alpha^5 & \alpha^2 & \infty \end{array} .$$

III - Extended codes over the extension field $\mathbb{F}_{q^m}$.

Let $L = \{\alpha_1,\dots,\alpha_n\} \subset \mathbb{F}_{q^m}$ and let g a polynomial with degree r, over $\mathbb{F}_{q^m}$, with no roots in L.

Denote $\overline{\Gamma}(L,g)$ the set of n-tuples $(x_{\alpha_1},\dots,x_{\alpha_n}) \in \mathbf{F}^n_{q^m}$ such that

$$\sum_{\alpha_i \in L} \frac{x_{\alpha_i}}{z-\alpha_i} = 0 \bmod g.$$

It is well known (cf. Theorem 4, p. 340 of [6]) that $\overline{\Gamma}(L,g)$ is a generalized Reed Solomon codes $GRS_{n-r}(L,v)$ where $v = (v_1,\dots,v_n)$ is defined by

$v_i = \dfrac{g(\alpha_i)}{L'(\alpha_i)}$ $(i = 1,\dots,n)$, $(L(X) = \prod_{i=1}^{n} (X-\alpha_i)$, $L'(X)$ is the formal derivative of $L(X)$).

Any codeword is of the following form :

$$C_u = (v_1 u(\alpha_1),\dots,v_n u(\alpha_n))$$

where u is a polynomial in $\mathbf{F}_{q^m}[X]$ with degree < n-r.

Lemma 4 : For any $\sigma \in PGL(2,q^m)$ we have

$$\overline{\Gamma}(L,g) = \overline{\Gamma}(\sigma L,\sigma^{-1}g).$$

Proof : Use a parity check matrice of the form

$(\alpha_i^t\, g(\alpha_i)^{-1})$ $\quad (0 \le t \le r,\ 1 \le i \le n)$, and verify that

$$\frac{(\sigma\alpha_i)^t C_{\alpha_i}}{\sigma^{-1} g\sigma\alpha_i} = p(\alpha_i)\, \frac{C_{\alpha_i}}{g(\alpha_i)}$$

where $p(x)$ is some polynomial with degree $\le r$.

In the following we suppose $n-r \ge 2$.

Lemma 5 : $\overline{\Gamma}(L,g) = \overline{\Gamma}(L,g')$ if and only if

$$g' = \lambda g \text{ for some } \lambda \in \mathbf{F}_{q^m}\setminus\{0\}.$$

Proof ($\Leftarrow$) : clear

($\Rightarrow$) consider the polynomial $U(x) = x - \alpha_i$ $\quad (\alpha_i \in L)$.
The codeword $C_u = (v_1 u(\alpha_1), \ldots, v_n u(\alpha_n))$
has a single 0 in coordinate α_i. Since $\bar{\Gamma}(L,g) = \bar{\Gamma}(L,g')$ there exists $u'(x) = \lambda(x-\alpha_i)$ $(\lambda \neq 0)$ such that :

$$C_{u'} = (v_1' u'(\alpha_1), \ldots, v_n' u'(\alpha_n)) = C_u$$

This implies : $v_j u(\alpha_j) = v_j' u'(\alpha_j)$ for $j = 1, \ldots, n$.
That is : $g(\alpha_j) - \lambda g'(\alpha_j) = 0$ for any $\alpha_j \in L \backslash \{\alpha_i\}$.
But polynomial $g(x) - \lambda g'(x)$ has degree at most r. □

Theorem 3 : $\bar{\Gamma}(L,g) = \bar{\Gamma}(L',g')$ if and only if there exists $\sigma \in PGL(2,q^m)$ such that $\sigma L = L'$ and $\sigma^{-1} g \sim g'$.

Proof ($\Leftarrow$) : clear by lemma 4.

($\Rightarrow$) Let $L = \{\alpha_1, \ldots, \alpha_n\}$ and $L' = \{\alpha_1', \ldots, \alpha_n'\}$.

As $PGL(2,q^m)$ is 3-transitive, there exists $\sigma \in PGL(2,q^m)$ such that $\alpha_i' = \sigma(\alpha_i)$ $(i = 1,2,3)$.
We have to show that $\alpha_i' = \sigma(\alpha_i)$ for $i = 4, \ldots, n$.
By lemma 4, we many suppose $L' = \{\alpha_1, \alpha_2, \alpha_3, \alpha_4', \ldots, \alpha_n'\}$ and now, we have to prove that $\alpha_i' = \alpha_i$ for $i = 4, \ldots, n$. Suppose there exists ℓ $(\ell \geq 4)$ such that $\alpha_\ell' \neq \alpha_\ell$.

Let $h(x) = x - \alpha_\ell$ and $w(x) = x - \alpha_t$ with $t \notin \{1,2,3,\ell\}$.
then $C_h = (v_1 h(\alpha_1), v_2 h(\alpha_2), \ldots, v_n h(\alpha_n))$ and
$C_w = (v_1 w(\alpha_1), v_2 w(\alpha_2), \ldots, v_n w(\alpha_n))$
are two codewords in $\bar{\Gamma}(L,g)$.

h_w has a single 0 in coordinate α_ℓ, and C_w has a single 0 in coordinate α_t.
So, as $\bar{\Gamma}(L,g) = \bar{\Gamma}(L',g') = GRS_{n-r}(L',v')$ there exists $h'(x) = \lambda(x-\alpha_\ell')$ and $w'(x) = \mu(x-\alpha_t')$ forme some λ, μ in $\mathbf{F}_{q^m} \backslash \{0\}$ such that :

$$C_{h'} = (v_1'h'(\alpha_1), v_4'h'(\alpha_2), v_3'h'(\alpha_3), \ldots, v_n'h'(\alpha_n')) = C_h$$

and

$$C_{w'} = (v_1'w'(\alpha_1), \ldots\ldots\ldots\ldots, v_n'w'(\alpha_n')) = C_w.$$

Therefore $v_i'h'(\alpha_i) = v_ih(\alpha_i)$ and $v_i'w'(\alpha_i) = v_iw(\alpha_i)$ for $i = 1,2,3$.

So we have :

$$\frac{h(\alpha_i)}{w(\alpha_i)} - \frac{h'(\alpha_i)}{w'(\alpha_i)} = 0 \text{ for } i = 1,2,3.$$

But

$$\frac{h(x)}{w(x)} - \frac{h'(x)}{w'(x)} = \frac{P(x)}{w(x)w'(x)} \quad ,$$

where $P(x)$ is not zero and with degree 2. A contradiction since, $P(\alpha_i) = 0$ for $i = 1,2,3$.

We may conclude by lemma 5. □

Theorem 4 : An extended code $\bar{\Gamma}(L,g)$ over $\mathbb{F}_{q^m}$ where $L = \mathbb{F}_{q^m} \cup \{\infty\}$ is a cyclic code if and only if $g = (z-\beta_1)^a(z-\beta_2)^a$ where β_1 and β_2 are two conjugate elements in $\mathbb{F}_{q^{2m}} \setminus \mathbb{F}_{q^m}$ and where a is an integer ≥ 1.

Proof : By the same arguments as in the proof of theorem 1 we have $\bar{\Gamma}(L,g) = \bar{\Gamma}(L',g')$ where $g' = (z-\beta_1')^a(z-\beta_2')^a$ (β_1' and β_2' two conjugate elements in $\mathbb{F}_{q^{2m}} \setminus \mathbb{F}_{q^m}$).

Theorem 4 is now an application of theorem 3.

Remark : Algebraic geometric arguments are used by Stichtenoch in [7] to show the main results of this section.

Acknowledgement

The author wishes to thank the referees for their careful reading, comments, and suggestions which have helped greatly in improving the presentation of these results.

References :

[0] E.R. Berlekamp and O. Moreno. "Extended double error correcting binary Goppa Codes are cyclic".
IEEE Trans. Inform. Theory, Vol. II.19, pp. 817-818, nov. 1973.

[1] V.D. Goppa. "A new class of linear error correcting codes". Prob. Peredach Inform. Vol. 6 n° 3, pp. 24-30, Sept. 1970.

[2] O. Moreno. "Symmetries of Binary Goppa Codes".
IEEE Trans. Inform. Theory, Vol. II.25, n° 5, pp. 609-612, sept. 1979.

[3] K.K. Tzeng and K. Zimmermann. "On extending Goppa Codes to cyclic Codes".
IEEE Trans. Inform. Theory, Vol. II.21, pp. 712-716, nov. 1975.

[4] K.K. Tzeng and Chie Y. Yu. "Characterization theorems for extending Goppa Codes to cyclic codes".
IEEE Trans. Inform. Theory, Vol. II.25, n° 2, pp. 246-249, March 1979.

[5] A.L. Vishnevetskii. "Cyclicity of extended Goppa Codes".
Prob. Peredachi Inform., Vol. n° 3, pp. 14-18, sept. 1982.

[6] F.J. Mac Williams, N.J.A. Sloane. "The Theory of Error Correcting Codes".
North-Holland, 1978.

[7] H. Stichtenoth. "Geometric Goppa Codes of genus 0 and then Automorphism Groups". Preprint.
Fachbereich 6 - Mathematik
Universität - Gesamthochschule Essen

SOME BOUNDS FOR THE CONSTRUCTION OF GRÖBNER BASES

Volker Weispfenning
Mathematisches Institut der Universität
D-6900 Heidelberg , FRG

ABSTRACT. Let $R = K[X_1, \ldots, X_n]$ be a polynomial ring over a field. For any finite subset F of R, we put $m = \mid F \mid$, $d = max(deg(F) : f \in F)$, and we let s be the maximal size of the coefficients of all $f \in F$. $G = GB(F)$ denotes the unique reduced Gröbner basis for the ideal (F) (see [B3]). We show that the number $m' = \mid G \mid$ of polynomials in G and their maximal degree d' as well as the length of the computation of G from F (with unit cost operations in K) are bounded recursively in (n, m, d). The same applies to the degrees of the polynomials occuring during the computation. Moreover, for fixed (n, m, d), G can be computed from F in polynomial time and linear space, when the operations of K can be performed in polynomial time and linear space; in addition, the vector space dimension of the residue ring $R/(F)$ is computably stable under variation of the coefficients of polynomials in F. Corresponding facts hold for polynomial rings over commutative regular rings (see [We']) and non-commutative polynomial rings of solvable type over fields (see [KRW]). Our method does not apply to polynomial rings over Z or other Euclidean rings; in fact, we show that over Z, the length of the computation of G from F with unit cost operations in Z *does* depend on s.

INTRODUCTION. The success of Buchberger's method for computing Gröbner bases in the algorithmic theory of polynomial ideals has aroused a strong interest in the complexity of the method (see [B1], [Gi], [Gi'], [Hu], [La], [MaMe], [MM], [Wi]): Most of the results concern upper bounds on the degrees of the polynomials in a reduced Gröbner basis for the case of two variables ([B1], [Gi'], [La]), three variables ([Wi], [MM]), or special assumptions on the ideals considered ([Gi'], [La], [MM]); [Gi] provides a bound on these degrees for arbitrary homogeneous polynomials. In [B1], [Wi] these bounds concern also all polynomials arising during the Gröbner basis calculation. [B1] provides in addition a tight bound for the number of polynomials in a reduced Gröbner basis for the bivariate case. Worst case lower bounds are obtained in [MaMe], [B1], [MM], [Hu].

Let K be an arbitrary field, let R be the polynomial ring $K[X_1, \ldots, X_n]$, let $\prec$ be an admissible linear ordering of the set T of terms in R and let F be a set of m polynomials of total degree at most d in R. $G = GB(F)$ denotes the unique reduced Gröbner basis for the ideal $I = (F)$ generated by F in R (see [B3]). We assume that the elements a of K are presented as words over a fixed finite alphabet in such a way that the field operations in K and tests $(a = 0)$? can be performed in polynomial time and linear space. (This assumption is satisfied in most of the fields studied in computer algebra.) We let $s(a)$ denote the *size* of a, i.e. the length of a word representing the field element a.

In this note, we show that the following items are bounded by functions of n, m, d: The number $\mid G \mid$ of polynomials in G, the maximal degree of all polynomials occuring in a computation of G from F, and the number of steps required for such a computation, when the field operations in K cost unit time. The bounds are independent of the field K. From the proof presented here, they seem to depend on the choice of the admissible ordering $\prec$ of T; in fact, however, this dependence can be eliminated by an application of König's tree lemma. This and further consequences of our method will be presented in [We"]. We have no explicit description of the bounds; their existence is proved by an application of the compactness theorem of first-order logic (see [Ke]). Nevertheless, we can show (using the decidability of the theory of algebraically closed fields) that the bounds depend recursively on n, m, d. Moreover, we find that *for fixed* n, m, d, the size $s(G)$ of the coefficients of the polynomials in G is linear in the corresponding coefficient size $s(F)$ for F, and that G can be computed from F in polynomial time and linear space, when the field operations in K can be performed in polynomial time and linear space. The same applies to the finitely many vector space dimensions of $R/(F)$ obtainable by varying the coefficients of the polynomials in F. In addition, these dimensions are completely determined by the zeroes of a finite computable system of polynomials.

The crux of the argument is the fact that the single steps in the computation of G can be coded by first–order formulas in K, and that the class of fields is axiomatized by first–order axioms. Thus corresponding results are valid in related situations, including polynomial rings over commutative regular rings (see [We']) and non–commutative polynomial rings of solvable type over fields (see [KRW]). The method fails for polynomial rings over the integers (see [B2] and the references given there), and more generally for polynomial rings over Euclidean rings (see [KRK]). Here, the termination of the algorithm depends not only on Dickson's lemma, but also on the termination of the Euclidean algorithm in the ground ring , which may fail in non–standard models of this ring. In fact, we prove that over the ground ring Z of integers, the length of the computation of G from F *does* depend on $s(F)$, even if the ring cost only unit time.

I am indebted to *R. Loos*, who made me aware of the questions studied in this note.

1. CODING GRÖBNER BASIS CONSTRUCTIONS IN FIELD THEORY

We consider formulas of the elementary theory of fields. They are obtained inductively from equations $f(x_1, \ldots, x_m) = g(x_1, \ldots, x_m)$, where f, g are polynomials with integer coefficients in some variables x_i, by means of $\neg$ (*negation*), $\wedge$ (*conjunction*) , $\vee$ (*disjunction*), and *quantification* $\exists x, \forall x$ over some variables. A formula is *quantifier–free* (*q.f.*) if it contains no quantifier $\exists x, \forall x$. Our goal is to express the construction of a reduced Gröbner basis G from a given ideal basis F (see [B3]) by q.f. formulas, provided this construction is bounded in 'size' in a suitable way.

Let n, m, d be positive integers, let K be a field, $R = K[X_1, \ldots, X_n]$ a polynomial ring over K, T the set of *terms* (power–products of the X_i) in R, and let $\prec$ be an admissible ordering of T. Then any polynomial $f \in R$ of (total) degree $\leq d$ can be recovered uniquely from its sequence $\mathbf{c} = \mathbf{c}(f) = (c_1, \ldots, c_s)$ of coefficients. (We regard f as polynomial of formal degree d in dense representation, adding zero coefficients as necessary; the coefficients are ordered in the decreasing order of their associated terms. So $s = \binom{d+n}{n}$, and c_1 is the coefficient ent of the *head–term HT(f)* of F, provided $c_1 \neq 0$.) A sequence $F = (f_1, \ldots, f_m)$ of polynomials in R of degrees $\leq d$ can then be recovered from d and the concatenated sequence $\mathbf{c}(F) = \mathbf{c}(f_1) * \ldots * \mathbf{c}(f_m)$.

Let us now consider a computation leading from a finite sequence F of polynomials in R to a reduced Gröbner basis $G = GB(F)$ for the ideal $I = (F)$ generated by F in R. Disregarding the control structure of the computation, we may view it as a finite sequence $F = F_0 \mapsto F_1 \mapsto \ldots \mapsto F_s = G$ of finite sequences F of polynomials in R, such that each $H' = F_{k+1}$ is obtained from its predecessor $H = F_k$ by one of the following steps $S_1, \ldots, S_4$:

Let $H = (f_1, \ldots, f_m)$.

(S_1) $H' = (f_1, \ldots, f_m, h)$, where $h = SPol(f_i, f_j)$, $1 \leq i \prec j \leq m$, is the *S-polynomial* of two members of H.

(S_2) $H' = (f_1, \ldots, f_{i-1}, f_i', f_{i+1}, \ldots, f_m)$, where f_i, $1 \leq i \leq m$, is a member of H, and f_i' is obtained from f_i by a 1-step reduction using the polynomials $\{f_1, \ldots, f_{i-1}, f_{i+1}, \ldots, f_m\}$.

(S_3) $H' = (f_1, \dots, f_{i-1}, \tilde{f}_i, f_{i+1}, \dots, f_m)$, where f_i, $1 \le i \le m$, is a member of H, and $\tilde{f}_i = HT(f_i)^{-1} \cdot f_i$ is the monic polynomial corresponding to f_i.

(S_4) $H' = (f_1, \dots, f_{i-1}, f_{i+1}, \dots, f_m)$, where f_i, $1 \le i \le m$, is a zero polynomial in H.

Moreover, F is a Gröbner basis iff for every step $F = F_0 \mapsto F_1$ of type 1, there exists a finite sequence of steps of type 2 constituting a reduction $h \xrightarrow{*} 0$ of h to zero using polynomials in F. Let us call F an *s-step Gröbner basis*, if every such reduction to zero of an S-polynomial of polynomials in F can be performed in at most s steps.

Assume now that we have fixed n, $m = | H |$, and a bound on the degrees of all polynomials in H. Then the degree of all polynomials in H' can be bounded by $2d$, if H' is obtained from H by step S_1 or S_2 , and by d, if H' is obtained from H by step S_3 or S_4. Moreover, the number $m' = | H' |$ of polynomials in H' is $m + 1$ in the first case, m in case 2 and 3, and $m - 1$ in case 4.

As a consequence, we have the following result.

LEMMA 1.1 For all positive integers n, m, d and $1 \le i \le 4$, one can compute q.f. formulas $\varphi_{i,n,m,d}(x_1, \dots, x_r, x'_1, \dots, x'_{r'_i})$, where $r = m \cdot \binom{d+n}{n}$, $r'_1 = \binom{d'_1+n}{n} \cdot (m+1)$, $r'_2 = \binom{d'_2+n}{n} \cdot m$, $r'_3 = \binom{d'_3+n}{n} \cdot m$, $r'_4 = \binom{d'_4+n}{n} \cdot (m-1)$, $d'_1 = d'_2 = 2d$, $d'_3 = d'_4 = d$, such that for every field K, and all sequences $H = (f_1, \dots, f_m)$, $H'_i = (g_1, \dots, g_{m'_i})$ in $R = K[X_1, \dots, X_n]$ with $m'_1 = (m+1)$, $m'_2 = m'_3 = m'_4 = (m-1)$, $deg(f_j) \le d$, $deg(g_j) \le d'_i$, the following holds:

(1) Let $(c_1, \dots, c_r) = \mathbf{c}(H)$, $(c'_1, \dots, c'_{r'_i}) = \mathbf{c}(H'_i)$. Then for $i = 1, 2, 3, 4$,
$\varphi_{i,n,m,d}(c_1, \dots, c_r, c'_1, \dots, c'_{r'_i})$ holds in K iff H'_i is obtained from H by step S_i .

(2) Moreover, each $\varphi_{i,n,m,d}(\mathbf{x}, \mathbf{x}')$ is of the following form

$$\bigvee_{j \in J_i} [\psi_{i,j,n,m,d}(\mathbf{x}) \;\wedge \bigwedge_{h=1\dots r} [x'_h \cdot q_{h,i,j,n,m,d}(\mathbf{x}) = p_{h,i,j,n,m,d}(\mathbf{x}) \;\wedge\; q_{h,i,j,n,m,d}(\mathbf{x}) \neq 0]]$$

, where J_i is a finite set, $\psi_{i,j,n,m,d}(\mathbf{x})$ is a q.f. formula containing no x'_h, and $p_{h,i,j,n,m,d}(\mathbf{x})$, $q_{h,i,j,n,m,d}(\mathbf{x})$ are polynomials in $\mathbf{x}$ with integer coefficients.

The *construction of these formulas* is tedious, but straightforward. We illustrate the method for the case $m = 2$, $H = (f_1, f_2)$, and step S_1. Let $s = \binom{d+n}{n}$, $s' = \binom{2d+n}{n}$, $r = 2s$, $r' = 3s'$, $\mathbf{c}(H) = (c_1, \dots, c_r)$, $\mathbf{c}(H') = (c'_1, \dots, c'_{r'})$. Let $(t_1, \dots, t_s)[(t'_1, \dots, t'_{s'})]$ be all terms of degree $\le d$ [of degree $\le 2d$] in decreasing order. Then there exists a function

$$\alpha : \{1, \dots, s\}^2 \longrightarrow \{1, \dots, s'\}$$

such that $t'_{\alpha(i,j)} = lcm(t_i, t_j)$, and functions

$$\beta, \gamma : \{1, \dots, s\}^3 \longrightarrow \{1, \dots, s'\}$$

such that

$$t'_{\beta(i,j,k)} = t'_{\alpha(i,j)} \cdot t_k / t_i,$$

$$t'_{\gamma(i,j,k)} = t'_{\alpha(i,j)} \cdot t_k / t_j.$$

Moreover, β and γ are injective in k for fixed i, j.

The q.f. formula $\varphi_{1,n,2,d}(x_1, \dots, x_r, x'_1, \dots, x'_{r'})$ can now be defined as follows :

$$\bigvee_{i,j=1\dots s} \Big(\bigwedge_{k=1\dots(i-1)} x_k = 0 \;\wedge \bigwedge_{k=1\dots(j-1)} x_{s+k} = 0 \;\wedge \bigwedge_{h=1\dots(s'-s)} x'_h = 0 \;\wedge$$

$$\bigwedge_{h=(s'-s+1)\dots s'} x'_h = x_{h-(s'-s)} \wedge \bigwedge_{h=s'\dots(2s'-s)} x'_h = 0 \wedge \bigwedge_{h=(2s'-s+1)\dots 2s'} x'_h = x_{h-2(s'-s)} \wedge$$

$$\bigwedge_{h\in K_1} x'_{2s'+h} = 0 \wedge \bigwedge_{h\in K_2, h=\beta(i,j,k)} x'_{2s'+h} = x_k \wedge$$

$$\bigwedge_{h\in K_3, h=\gamma(i,j,\ell)} x'_{ss'+h} \cdot x_{s+j} = -x_i \cdot x_\ell \wedge \bigwedge_{h\in K_4, h=\beta(i,j,k)=\gamma(i,j,\ell)} x'_{2s'+h} \cdot x_{s+j} = x_k \cdot x_{s+j} - x_i \cdot x_\ell \Big),$$

where $K = \{1,\dots,s'\}$, $K' = \{h \in K : \exists k \textit{ with } h = \beta(i,j,k)\}$, $K" = \{h \in K : \exists \ell \textit{ with } h = \gamma(i,j,\ell)\}$, $K_1 = K \setminus (K' \cup K")$, $K_2 = K' \setminus K"$, $K_3 = K" \setminus K'$, $K_4 = K' \cap K"$.

Next, we extend lemma 1.1 to computations involving a finite number of steps $S_1,\dots,S_4$: Let $\mathbf{i} = (i_1,\dots,i_\ell)$ be a finite sequence of numbers in $\{1,2,3,4\}$. Then we define $m'(\mathbf{i},m)$ and $d'(\mathbf{i},d)$ by recursion on the length ℓ of $\mathbf{i}$: For $\ell = 1$, $m'(\mathbf{i},m) = m'_i$ and $d'(\mathbf{i},d) = d'_i$ as defined in lemma 1.1 ; $m'((i_1,\dots,i_{\ell+1}),m) = m'(i_{\ell+1}, m'((i_1,\dots,i_\ell),m)$, $d'((i_1,\dots,i_{\ell+1}),m) = d'(i_{\ell+1}, d'((i_1,\dots,i_\ell),d))$. Finally, we define $r'(\mathbf{i},n,m,d) = \binom{d'(\mathbf{i},d)+n}{n} \cdot m'(\mathbf{i},m)$.

Using this notation, lemma 1.1 has a straightforward extension to computations involving finitely many steps of type $S_1,\dots,S_4$: If $\mathbf{i} = (i_1,\dots,i_\ell)$, then we let $S_\mathbf{i}$ denote the sequence of steps $(S_{i_1},\dots,S_{i_\ell})$.

LEMMA 1.2 For all positive integers n,m,d , and all finite sequences $\mathbf{i}$ of integers in $\{1,2,3,4\}$one can compute q.f. formulas $\varphi_{\mathbf{i},n,m,d}(x_1,\dots,x_r,x'_1,\dots,x'_{r'})$, where $r = m\cdot\binom{d+n}{n}, r' = r'(\mathbf{i},n,m,d)$, such that the following assertions hold :

(1) For every field K and all sequences $H = (f_1,\dots,f_m), H' = (g_1,\dots,g_{m'})$ of polynomials in $R = K[X_1,\dots,X_n]$ with $deg(f_j) \le d$, $deg(g_j) \le d'(\mathbf{i},d)$, $\varphi_{\mathbf{i},n,m,d}(\mathbf{c}(H),\mathbf{c}(H'))$ holds in K iff H' is obtained from H by a sequence of steps of type $S_\mathbf{i}$.

(2) Moreover, each $\varphi_{\mathbf{i},n,m,d}(\mathbf{x},\mathbf{x}')$ is of the form described in lemma 1.1 (2) , with i replaced by $\mathbf{i}$.

PROOF. Induction on the length of $\mathbf{i}$.

In a very similar manner, one can compute q.f. formulas characterizing reduced s-step Gröbner bases for any given s:

LEMMA 1.3 For all positive integers n,m,d,s one can compute a q.f. formula $\gamma_{n,m,d,s}(x_1,\dots,x_r)$, where $r = m\cdot\binom{d+n}{n}$, such that for all fields K and all m-tuples H of polynomials in $R = K[X_1,\dots,X_n]$ of degree $\le d$, $\gamma_{n,m,\alpha,s}(\mathbf{c}(H))$ is valid in K iff H is a reduced s-step Gröbnerbasis in R .

Finally, we may combine lemma 1.2 and 1.3 with the fact that for any given M-tuple F of polynomials in R there is a finite sequence $S_\mathbf{i}$ of steps leading from F to the unique reduced Gröbner basis G of (F) .

COROLLARY 1.4 For all positive integers n,m,d, the following *infinitary* formula is valid in all fields K :

$$\forall x_1 \dots x_r \Big(\bigvee_{1\le s\in N} \bigvee_{1\le \ell\in N} \bigvee_{\mathbf{i}\in\{1,2,3,4\}^\ell} \exists x'_1 \dots x'_{r'} \delta_{\mathbf{i},n,m,d,s}(\mathbf{x},\mathbf{x}') \Big)$$

where $r = n \cdot \binom{d+n}{n}$, $r' = r'(\mathbf{i}, n, m, d)$, and $\delta_{\mathbf{i},n,m,d,s}(\mathbf{x}, \mathbf{x}')$ is the formula

$$\varphi_{\mathbf{i},n,m,d}(\mathbf{x}, \mathbf{x}') \ \wedge \ \gamma_{n,m'(\mathbf{i},m),d'(\mathbf{i},d),s)}(\mathbf{x}').$$

2. MAIN RESULTS

The compactness theorem of first–order logic (see [Ke]), rephrased for the elementary theory of fields, says the following : Let Φ be an (infinite) set of formulas such that the (infinitary) conjunction $\bigwedge \Phi$ over all formulas in Φ is unsatisfiable in any field K. Then there exists a finite subset Φ' of Φ such that the conjunction $\bigwedge \Phi'$ is unsatisfiable in any field K.

Using the notation of the previous section, we apply this theorem to the following set of formulas Φ :

$$\left\{ \neg \exists x'_1 \ldots x'_{r'} \delta_{\mathbf{i},n,m,d,s}(\mathbf{x}, \mathbf{x}') \ : \ 1 \leq s \in \mathbf{N}, \ 1 \leq \ell \in \mathbf{N}, \ \mathbf{i} \in \{1,2,3,4,\}^{\ell} \right\}$$

By corollary 1.4, $\bigwedge \Phi$ is unsatifiable in any field K, and so there exists a finite subset Φ' of Φ such that $\bigwedge \Phi'$ is unsatisfiable in any fields K. Recalling the definition of $\delta_{\mathbf{i},n,m,d,s}(\mathbf{x}, \mathbf{x}')$, we see that for $s \leq s'$, $\delta_{\mathbf{i},n,m,d,s}$ implies $\delta_{\mathbf{i},n,m,d,s'}$. So we obtain :

THEOREM 2.1 For all positive integers n, m, d there exists an integer s, a finite set I of finite sequences $\mathbf{i}$ of numbers in $\{1,2,3,4\}$, and for each $\mathbf{i} \in I$ a finite index set $J(\mathbf{i})$, q.f. formulas $\varepsilon_{\mathbf{i},j,n,m,d}(x_1, \ldots, x_r)$ and rational functions $p_{h,\mathbf{i},j,n,m,d}(x_1, \ldots, x_r) \ / \ q_{h,\mathbf{i},j,n,m,d}(x_1, \ldots, x_r)$ $(1 \leq h \leq r'(\mathbf{i}, n, m, d), \ j \in J(\mathbf{i}), \ r = m \cdot \binom{d+n}{n})$ such that the following assertions hold :

(1) For every field K and every m-tuple F of polynomials in $R = K[X_1, \ldots, X_n]$ of degree $\leq d$, there exists $\mathbf{i} \in I$ and $j \in J(\mathbf{i})$, such that $\varepsilon_{\mathbf{i},j,n,m,d}(\mathbf{c}(F))$ holds in K .

(2) Whenever for K and f as in (1), $\varepsilon_{\mathbf{i},j,n,m,d}(\mathbf{c}(F))$ holds in K, then :

2.1 The reduced Gröbner basis G of (F) is obtained from F by a sequence of steps of type $S_{\mathbf{i}}$.

2.2 G has $m'(\mathbf{i}, m)$ elements, each of degree $\leq d'(\mathbf{i}, d)$, and G is an s-step Gröbner basis.

2.3 Let $\mathbf{c}(G) = (c'_1, \ldots, c'_{r'})$ with $r' = r'(\mathbf{i}, n, m, d)$. Then for $1 \leq h \leq r'$, $c'_h = p_{h,\mathbf{i},j,n,m,d}(\mathbf{c}(F)) \ / \ q_{h,\mathbf{i},j,n,m,d}(\mathbf{c}(F))$, and $q_{h,\mathbf{i},j,n,m,d}(\mathbf{c}(F)) \neq 0$ in K .

Call a field K a *POLYLIN field* (with respect to a fixed representation of the elements of K by words over a finite alphabet) , if the field operations $=, -, \cdot, /$ and tests $(a = 0)$? can be performed in polynomial time and linear space. Most fields considered in computer algebra such as the rationals, finite algebraic number fields, finite fields, and rational function fields in finitely many indeterminates over these fields are POLYLIN fields (see [CA]). An induction on the length of q.f. formulas shows easily that in a POLYLIN field K, any q.f. formula ρ can be evaluated in polynomial time and linear space.

COROLLARY 2.2 Let n, m, d be fixed positive integers, let K be a POLYLIN field, $R = K[X_1, \ldots, X_n]$, and let F be an m-tuple of polynomials in R of degree $\leq d$. Then the reduced Gröbner basis G of (F) can be computed from $\mathbf{c}(F)$ in polynomial time and linear space. In particular, the size of $\mathbf{c}(G)$ is linearly bounded by the size of $\mathbf{c}(F)$.

PROOF. First, one evaluates the q.f. formulas $\varepsilon_{\mathbf{i},j,n,m,d}(\mathbf{c}(F)$ until one finds $\mathbf{i} \in I$, $j \in J(\mathbf{i})$ such that the corresponding formula holds in K. Then one evaluates the corresponding rational functions $p_{h,\mathbf{i},j,n,m,d}(\mathbf{c}(F)) \ / \ q_{h,\mathbf{i},j,n,m,d}(\mathbf{c}(F))$. By the hypothesis, both can be done in polynomial time and linear space.

The corollary leads to a somewhat paradoxical situation : It proves the *existence* of a POLY-LIN–algorithm without really exhibiting such an algorithm. (Notice that each formula $\varepsilon_{\mathbf{i},j,n,m,d}(\mathbf{x})$ as well as each of the rational functions $p_{h,\mathbf{i},j,n,m,d}(\mathbf{x})$ / $q_{h,\mathbf{i},j,n,m,d}(\mathbf{x})$ is constructed explicitly; the finite set I , however, was found non–constructively.) Nevertheless, we have the following result :

COROLLARY 2.3 The integers s, the finite sets I and $J(\mathbf{i})$ in theorem 2.1, and hence the bounds $m' = max(m'(\mathbf{i}, m) : \mathbf{i} \in I)$, $d' = max(d'(\mathbf{i}, d) : \mathbf{i}, \in I)$ depend recursively on n, m, d.

PROOF. By a slight modification of 1.4, one can avoid the existential quantifiers in the infinitary formula displayed, and instead arrive at an equivalent infinitary formula of the form

$$\forall x_1 \ldots x_r \Big(\bigvee_{1 \le s \in \mathbf{N}} \; \bigvee_{1 \le \ell \in \mathbf{N}} \; \bigvee_{\mathbf{i} \in \{1,2,3,4\}^\ell} \delta'_{\mathbf{i},n,m,d,s}(\mathbf{x}) \Big)$$

true in all fields. In order to test the validity of the finite approximations

$$\forall x_1 \ldots x_r \Big(\bigvee_{\mathbf{i} \in I} \delta'_{\mathbf{i},n,m,d,s}(\mathbf{x}) \Big), \quad (I \text{ finite})$$

in all fields, one may employ a decision method for the theory of algebraically closed fields (see e.g. [Ra]). until for some s and I one gets a positive answer, which is guaranteed to happen eventually by the compactness theorem.

We note another consequence of theorem 2.1 on the stability of the vector space dimension of the residue ring R/I under variation of coefficients.

COROLLARY 2.4 For any triple (n, m, d) one can compute a sequence of polynomials $h_i(x_1, \ldots, x_r)$ $(r = m \cdot \binom{d+n}{n}$, $1 \le i \le \ell)$, such that for all fields K , $R = K[X_1, \ldots, X_n]$, and all sequences $F = (f_1, \ldots, f_m)$ of polynomials in R of degree $\le d$, the dimension of $R/(F)$ over K is determined by the values of $sgn(h_i(\mathbf{c}(F))$ for $1 \le i \le \ell)$.

PROOF. The sequence $(h_1, \ldots, h_\ell)$ consists of all polynomials occuring in the formulas $\varepsilon_{\ldots}(\mathbf{x})$ together with the polynomials $p_{\ldots}(\mathbf{x})$ and $q_{\ldots}(\mathbf{x})$ appearing in theorem 2.1. The vanishing or non-vanishing of these polynomials determines the highest terms of the reduced Gröbner basis of (F) and thus by [B3]the dimension of $R/(F)$ over K.

Finally, we remark that our results (except possibly 2.3 and 2.4) hold — mutatis mutandis — for all Gröbner basis constructions, where the ground rings are described by elementary axioms, and the single steps of the construction as well as the characterization of an s-step Gröbner basis can be expressed by formulas about the ground ring. This is the case e.g. for Gröbner basis constructions in the non–commutative polynomial rings ofsolvable type considered in [KRW], and in polynomial rings over commutative regular rings (see [We']). Moreover, in these cases corollary 2.3 remains valid (use [We] for the second case). In the first case, corollary 2.4 holds for left ideals.

As pointed out in the introduction, our method does not apply to polynomial rings over Z or other Euclidean rings. We are going to show now that theorem 2.1 fails in $R = Z[X_1, \ldots, X_n]$: Let $e(k)$ be the *k.th–Fibonacci number* $(e(0) = 0,\ e(1) = 1,\ e(k+2) = e(k) + e(k+1))$. Then the Euclidean algorithm for computing $gcd(e(k+1), e(k)) = 1$ takes exactly k steps for $k \ge 1$. Accordingly, the polynomials $e(k+1)X$, $e(k)X$ are mutually reducible in the sense of [KRK] k times for $k \ge 1$. Since k is not bounded by $n = 1,\ m = 2,\ d = 1$, this contradicts theorem 2.1.

REFERENCES

[B1] B. Buchberger, A note on the complexity of constructing Gröbner bases, in [EC83], pp. 137–134.

[B2] B.Buchberger, A critical-pair/completion algorithm for finitely generated ideals in rings, in *Logic and Machines: Decision Problems and Complexity*, Münster 1983, Springer LNCS vol. 171.

[B3] B.Buchberger, Gröbner bases: An algorithmic method in polynomial ideal theory, chap. 6 in *Recent Trends in Multidimensional System Theory*, N. K. Bose Ed., Reidel Publ. Comp., 1985.

[CA] *Computer Algebra, Symbolic and Algebraic Computation*, B.Buchberger, G. E. Collins, R. Loos Eds., Springer Verlag 1982/83.

[EC83] *Computer Algebra, EUROCAL '83* , J. A. van Hulzen Ed., Springer LNCS vol. 162.

[EC85] *EUROCAL '85*, B.F.Caviness Ed., Springer LNCS vol. 204.

[ES84] *EUROSAM '84*, J. Fitch Ed., Springer LNCS vol. 174.

[Gi] M. Giusti, Some effectivity problems in polynomial ideal theory, in [ES84], pp. 159–171.

[Gi'] A note on the complexity of constructing standard bases, in [EC85], pp.411–412.

[Hu] D. T. Huynh, The complexity of the membership problem for two subclasses of polynomial ideals, SIAM J. of Comp. 15 (1986), 581–594.

[KRK] A. Kandri-Rody, D.Kapur, Algorithms for computing Gröbner bases of polynomial ideals over various Euclidean rings, in [ES84], pp. 195–205.

[KRW] A. Kandri-Rody, V.Weispfenning, Non–commutative Gröbner bases in algebras of solvable type, 1986, submitted.

[Ke] J. Keisler, Fundamentals of model theory, in *Handbook of Mathematical Logic*, J. Barwise Ed., North–Holland, 1977.

[La] D. Lazard, Gröbner bases, Gaussian elimination and resolution of systems of algebraic equations, in [EC83], pp. 146–156.

[MaMe] E. W.Mayr, A. R.Meyer, The complexity of the word problem for commutative semigroups and polynomial ideals, Adv. math. 46 (1982), 305–329.

[MM] H. M. Möller, F. Mora, Upper and lower bounds for the degree of Gröbner bases, in [ES84], pp 172–183.

[Ra] M. O. Rabin, Decidable theories, in *Handbook of mathematical Logic*, J. Barwise Ed., North–Holland, 1977.

[We] V. Weispfenning, Model–completeness and elimination of quantifiers for subdirect products of structures, J. of Algebra, 36 (1975), 252–277.

[We'] V. Weispfenning, Gröbner bases in polynomial rings over commutative regular rings, to appear in Proc. EUROCAL '87, Leipzig, Springer LNCS .

[We"] V. Weispfenning, Constructing universal Gröbner bases, 1987, submitted.

[Wi] F. Winkler, On the complexity of the Gröbner basis algorithm over $K[x, y, z]$, in [ES84], pp. 184–194.

ANALYSIS OF A CLASS OF ALGORITHMS FOR PROBLEMS ON TRACE LANGUAGES

A. Bertoni - M. Goldwurm - N. Sabadini
Dipartimento di Scienze dell'Informazione - Università di Milano

Abstract The time complexity of a class of algorithms for problems on trace languages is studied both in the worst and in the average case. Membership problem for regular trace languages and the problem of counting the cardinality of a trace can be solved by algorithms belonging to such a class. These algorithms are described by a recursive procedure scheme whose input is a trace represented by a corresponding string. Under reasonable assumptions the worst case time complexity on traces of length n is $\theta(n^{\alpha})$, where α is the cardinality of the largest clique in the concurrent alphabet $\langle\Sigma,C\rangle$.

The average case behaviour is studied under two different assumptions on the distribution of input set. In the first case equidistributed strings of length n are considered and it is proved that the average time complexity is $\theta(n^{k})$, where k is the number of connected components of the complementary alphabet $\langle\Sigma,C^{c}\rangle$. In the second case equally probable traces of length n are considered and it is proved that the average time complexity is $\theta(n^{h})$, where h is the multeplicity of the minimum modulus root of the polynomial $\Sigma_{j=0,\alpha}(-)^{j}\, r_j\, x^{j}$ and each r_j is the number of j-cliques in $\langle\Sigma,C\rangle$.

This research has been supported by Ministero della Pubblica Istruzione in the frame of the Project (40%) "Modelli e specifiche di sistemi concorrenti".

1. Introduction

The theory of trace languages, introduced by Mazurkiewicz [9] in order to describe the semantics of concurrent program schemes, is particularly interesting from a theoretical point of view. In fact, this approach can describe some properties of concurrent program schemes in the frame of formal language theory [1],[10],[7]; furthermore, from an algebraic point of view, trace languages are related to the theory of free partially commutative monoids developped by Foata and Cartier [4] in order to give an algebraic basis to the combinatorial theory of rearrangements.

In the following we recall some basic definitions on trace languages: a concurrent alphabet is a couple $<\Sigma,C>$, where Σ is a finite set of symbols and C is a simmetric, irreflexive relation on Σ; so, a concurrent alphabet can be viewed as a non oriented graph. The free partially commutative monoid (f.p.c.m.) $F(\Sigma,C)$ is the quotient monoid $\Sigma^*/=_C$, where $=_C$ is the least congruence on Σ^* which extends the commutativity axioms {ab=ba/ aCb }; by $[\ ]_C: \Sigma^* \dashrightarrow F(\Sigma,C)$ we denote the canonical morphism induced by the congruence $=_C$. A trace t is an element of $F(\Sigma,C)$ and a trace language T is a subset of $F(\Sigma,C)$; the length $|t|$ of a trace t is the length of every string x such that $[x]_C=t$, while the cardinality $\#t$ is the number of strings in t, interpreted as an equivalence class. A trace $t\in F(\Sigma,C)$ of lenght n can be represented in a synthetic way as a partially ordered set $<\{1,2,\ldots,n\},\ll>$ labelled by the symbols in t ([9]); more precisely, let t be the trace $[x_1x_2\ldots x_n]_C$, where $x_i\in\Sigma\ \forall i$, then we define the relation $\ll\ \subset \{1,2,\ldots,n\}^2$ as the reflexive and transitive closure of the relation R such that

iRj iff $(i<j$ and $\neg x_iCx_j)$; the vertex i is labelled by x_i.

Example Given $\Sigma = \{a,b,c,d\}$, $C = \{(a,b),(b,c)\}$, the trace $t = [abbcd]_C$ is, as equivalence class, the set {abbcd,abcbd,acbbd,babcd,bacbd,bbacd}, while the corresponding partial order is:

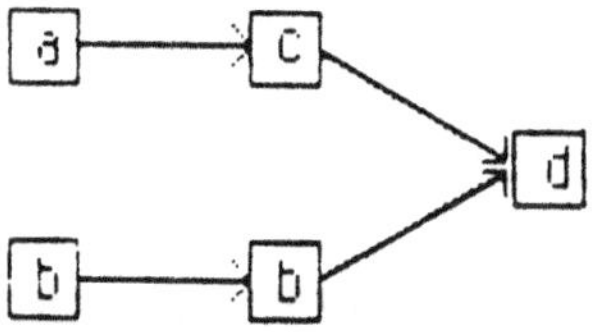

The class of regular (trace) languages on a f.p.c.m. $F(\Sigma,C)$ is the least class of languages containing all the finite languages and closed with respect to union, product,and closure. It is possible to prove [11] that an equivalent definition is the following: a trace language is regular iff there exists a regular language $L\subset\Sigma^*$ that generates T, that is $T = \{[x]_C\ /\ x\in L\}$;

in the following, we will denote with $[L]_C$ the trace language generated by L.

In this paper, we use some general algebraic techniques (in particular the theory of formal power series and generating functions) to state some properties of classes of algorithms on trace languages. In particular, we give a complete complexity analysis (worst case, average case) for the algorithm, presented in [2], that solves the membership problem for regular trace languages. This algorithm can be seen as a particular instance of a general procedure scheme that we introduce in the second section of this paper. It is easy to prove that, under reasonable assumptions on the basic operations of the procedure scheme, for inputs of size n the worst case time complexity W(n) is $\theta(n^{\alpha})$, where α is the clique-number of $\langle\Sigma,C\rangle$.

Moreover we study the average behaviour of the procedure scheme under different assumptions on the set of inputs. In the first case we consider equally probable strings of length n, and we prove that the average time complexity E(n) is $\theta(n^k)$, where k is the number of connected components of the complementary graph $\langle\Sigma,C^c\rangle$. In the second case we consider equally probable traces of length n. Under enough general assumptions, we prove that the average time $E_t(n)$ required by the procedure scheme is $\theta(n^h)$, where h is the multiplicity of the minimum modulus root of the polynomial $G(x) = \Sigma_{j=0,\alpha}(-)^j \Gamma_j x^j$ and each Γ_j is the number of j-cliques in $\langle\Sigma,C\rangle$.

In the last section these results are compared one to another on several examples, and some open problems are stated.

2. A general procedure scheme

The class of algorithms we are interested in is defined by the following procedure scheme :

```
Function P(t : trace)
 if |t|=1 then return f(t);
  else  begin
        S:= ∅ ; (S is a variable of type "set", ∅ is the empty set)
        for any t' such that ∃σ∈Σ and t'σ = t do
                 S:= (P(t'),σ) ∪ S;      ( σ∈Σ, t'∈F(Σ,C) )
        return Op(S);
   end
```

where f and Op are operation symbols. We can obtain particular algorithms from this procedure scheme, by interpreting these symbols as suitable operations.

We show two interesting examples:

1) Membership problem for Regular Trace Languages

Given a regular grammar $G=\langle V_N,\Sigma,P,q_0\rangle$ generating a language $L \subset \Sigma^*$, we consider the following problem:

Instance: $x \in \Sigma^*$;

Question: $[x]_C \in [L]_C$?

It is easy to show that the problem can be solved by checking whether q_0 belongs to U(t), where the set $U(t) \subset V_N$ is given by the following procedure :

```
Function  U(t : trace)
if |t|=1 then  return {q/(q→t) ∈ P};
 else    begin
           S:= ϕ ;
           for  t'σ = t  do  S := (U(t'),σ) ∪ S;
           return ∪_(A,σ)∈S {q/(q→q'σ)∈P, q'∈A};
         end
```

2) Cardinality of a trace

Given a concurrent alphabet $\langle\Sigma,C\rangle$, we consider the following problem :

Instance : $x \in \Sigma^*$;

Question : determine $\#([x]_C)$.

This problem can be solved by the following algorithm :

```
Function  #(t : trace)
if |t|=1 then  return 1;
 else    begin
           S:= ϕ ;
           for  t'σ = t  do  S := (#(t'),σ) ∪ S;
           return Σ_(N,σ)∈S N;
         end
```

Given these examples, we come back to the general procedure scheme. The general problem solved by the scheme is the following :

Instance : $x \in \Sigma^*$;

Question : compute $P([x]_C)$, where P is defined by the previous procedure.

Such a procedure is based on the recursive call of the function P applied to all the prefixes of the trace $t = [x]_C$, where a prefix of t is a trace t' such that $t=t' \cdot t''$ for a suitable trace t".

Let us denote the number of prefixes of a trace t by $\mathfrak{I}(t)$. Using an efficient representation of the prefixes of a trace (see [3]) and supposing that "f" and "Op" require constant time, it is possible to prove that the time complexity of the previous scheme (implemented on R.A.M.) is proportional to $\mathfrak{I}([x]_C)$ for any input x. In both examples the basic operations require constant time on R.A.M.. So $\theta(\mathfrak{I}([x]_C))$ is the time complexity of both algorithms on input $x \in \Sigma^*$.

3. Worst case analysis

By the previous remarks, the problem of evaluating the worst case time complexity of the procedure, when the functions "f" and "Op" require constant time, is reduce to the combinatorial problem of estimating $W(n)=Max_{|x|=n}\{\mathfrak{I}([x]_C)\}$.

Theorem 1 If the basic operations "f" and "Op" require constant time, then the time complexity of the procedure scheme P on inputs of size n is $\theta(n^\alpha)$ in the worst case, where α is the clique number in $\langle\Sigma,C\rangle$.

Proof. Since the prefixes of a trace t can be seen as order ideals of the corresponding poset $\langle t, \ll\rangle$, it is easy to show that $\mathfrak{I}(t)$ is exactly the number of antichains in such a poset. By observing that every antichain contains less than $\alpha+1$ elements, we can state that $\mathfrak{I}(t) \leq \theta(|t|^\alpha)$ and therefore $W(n) = \theta(n^\alpha)$.□

4. Average analysis A

In this section we study the average behaviour of the procedure P assuming that each string of length n is equiprobable. The problem is reduced to the estimation of $E(n) = \sum_{|x|=n}\mathfrak{I}([x]_C) / |\Sigma|^n$; clearly the main point is the evaluation of $f(n) = \sum_{|x|=n} \mathfrak{I}([x]_C)$.

In order to solve the problem, we first define a suitable language L for evaluating the generating function of f(n), and then we use the following proposition for estimating the asymptotic growth of f(n).

Lemma 1 Let $\{f_n\}$ be a real, positive sequence whose generating function is $f(x) = P(x)/Q(x)$, where $P(x)$ and $Q(x)$ are real coefficient polynomials. Moreover we suppose that $Q(x)$ has only one root b of least modulus. Then $f_n = \theta(n^{r-1}(1/b)^n)$, where r is the multeplicity of the root b in $Q(x)$.

Proof (outline). The function f(x) can be written as finite sum of suitable rational functions whose Taylor series can be easily estimated. Then the evaluation of f_n is obtained by comparing the main terms of such a developments. □

In the following, for every string x and for every trace t, we denote by |x| and |t| respectively the length of x and the length of a representative string of t. Moreover, for every set A, we ambiguously denote the cardinality of A by |A| : the meaning of | | will be clear from the context. Let us define a new concurrent alphabet $\langle\Sigma_1, C_1\rangle$ such that:

- $\Sigma_1 = \Sigma \cup \Sigma'$, where Σ' is an isomorphic copy of Σ : $\Sigma' = \{ \sigma' / \sigma \in \Sigma\}$;
- for every $\sigma, \tau \in \Sigma$ $\sigma C_1 \tau \Leftrightarrow \sigma C \tau$;

$$\sigma' C_1 \tau' \Leftrightarrow \sigma C \tau ;$$

$$\sigma' C_1 \tau \Leftrightarrow \sigma C_1 \tau' \Leftrightarrow \sigma C \tau .$$

Let us consider the trace language $T \subset F(\Sigma_1, C_1)$ such that

$$T = \{ t / t = t_1 \cdot t_2,\ t_1 \in F(\Sigma, C_1),\ t_2 \in F(\Sigma', C_1) \},$$

and let L be the set $\{ x / x \in (\Sigma \cup \Sigma')^*, [x]_{C1} \in T \}$. It is easy to verify that L is regular and satisfies the following proposition :

Proposition 1 For every integer n $|\{ z / z \in L,\ |z|=n \}| = f(n)$, where f(n) is the sequence defined above.

Proof. The language $L \subset (\Sigma \cup \Sigma')^*$ can be defined also as the set

$\{ z / z \in (\Sigma \cup \Sigma')^*$, z satisfies conditions 1), 2), 3) }, where:

1) $z = x_1 ... x_m y_1 ... y_q$, $m \geq 0$, $q \geq 0$,either $m \neq 0$ or $q \neq 0$;

2) $y_1 \in \Sigma'$, $\forall i$ $x_i \in \Sigma$;

3) defined $A = \{y_1,\dots,y_q\} \cap \Sigma$, $B_i = \{ y_j \,/\, j \le i \,,\, y_j \in \Sigma' \}$ $\forall i \le q$, we require that $\forall y_j \in A$, $y_j C_1 y$ for every $y' \in B_{j-1}$. In other words if $y_j \in \Sigma$ then it commutes with every symbol $y' \in \Sigma'$ which precedes y_j in z.

By using such a definition of L, it can be easily proved that there exists a one to one correspondence between the set $\{ z \,/\, z \in L, |z|=n \}$ and the set $\{(x,t) \,/\, x \in \Sigma^*, t \in F(\Sigma,C), |x|=n , [x]_C = t \cdot t' \}$. Therefore the thesis is a consequence of the following equalities :

$$f(n) = \sum_{|x|=n} \mathfrak{I}([x]) = \sum_t |\{ x \,/\, x \in \Sigma^* , t \in F(\Sigma,C), |x|=n , [x]_C = t \cdot t' \}| =$$
$$= |\{(x,t) \,/\, x \in \Sigma^*, t \in F(\Sigma,C), |x|=n , [x]_C = t \cdot t' \}| . \square$$

Now let us define the finite automaton $\langle M,q_0,\delta,F \rangle$, which recognizes the language L : $M = 2^\Sigma$, $q_0 = \Sigma$, $F = 2^\Sigma$,

$\forall A \in 2^\Sigma$, $\forall \sigma \in \Sigma$ $\delta(A,\sigma) = A$ if $\sigma \in A$, otherwise $\delta(A,\sigma)$ is undefined;

$\forall \sigma' \in \Sigma'$ $\delta(A,\sigma') = A \cap I(\sigma)$, where $I(\sigma) = \{ \sigma_1 / \sigma C \sigma_1 \}$.

Note that $\delta(A,\sigma') = A$ iff σ commutes with each element of A.

It is easy to prove that our language L is recognized by the automaton $\langle M,q_0,\delta,F \rangle$. So by proposition 1, we can state the following sentence:

Proposition 2 For every n f(n) is exactly the number of strings of length n recognized by $\langle M,q_0,\delta,F \rangle$.

Example Let us consider the partially commutative alphabet $\langle \{a,b\},\{(a,b)\} \rangle$. The corresponding automaton is represented by the following transition diagram :

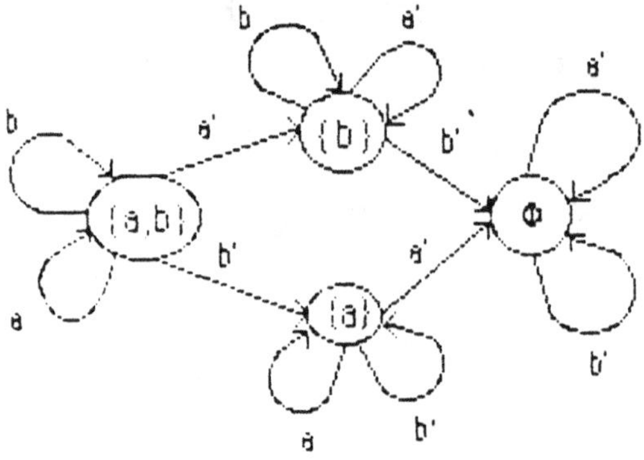

Using the transition matrix associated to such diagram, it is easily shown that the number of strins of length n accepted by the automaton is $2^{n-2}(n^2+3n+4)$. By proposition 2 such expression is equal to $\sum_{|x|=n} \mathfrak{I}([x])$.

In order to study the behaviour of $\{f(n)\}$ we list some definitions and properties of the automaton.

For every $A,A' \in 2^{\Sigma}$, $A \neq A'$, let us define:

- $l(A) = |\{\sigma / \ \sigma \in \Sigma \cup \Sigma', \delta(A,\sigma) = A\}|$;
- $\omega(A,A') = |\{\sigma' / \ \sigma' \in \Sigma', \delta(A,\sigma') = A'\}|$.

Moreover $\forall \sigma \in \Sigma$, let $I_C(\sigma)$ be the set $(I(\sigma))^C$.

Proposition 3 For every $A \in 2^{\Sigma}$, $l(A) \leq |\Sigma|$.

Proof. $l(A) = |A| + |\{\sigma / \ \sigma \in \Sigma', \delta(A,\sigma) = A\}| =$
$\leq |A| + |A^C| = |\Sigma|$.□

Proposition 4 For every $A \in 2^{\Sigma}$,
$l(A) = |\Sigma| \Leftrightarrow A^C$ is union of connected components of the graph $\langle\Sigma,C^C\rangle$.

Proof. By the inequality proved in proposition 3 we can state that

$$l(A) = |\Sigma| \Leftrightarrow A \subset \bigcap \{I(\sigma) / \sigma \in A^C\}.$$

We observe that $A \supset \bigcap \{I(\sigma) / \sigma \in A^C\}$: otherwise, there would be an element $a \in A^C$ such that, $\forall \sigma \in A^C$, $a \in I(\sigma)$, hence $a \in I(a)$ which is absurd for the irreflexivity of C. So we obtain :

$$l(A) = |\Sigma| \Leftrightarrow A = \bigcap \{I(\sigma) / \sigma \in A^C\}.$$

By definition of $I_C(\sigma)$, $A = \bigcap \{I(\sigma) / \sigma \in A^C\}$ holds iff

$$A^C = \bigcup \{I_C(\sigma) / \sigma \in A^C\}.$$

So, the last relation is true if and only if, $\forall a \in A^C$ and $\forall b \in \Sigma$, $(a,b) \in C^C$ ($b \in I_C(a)$) implies $b \in A^C$: that means that A^C is union of connected components of the graph $\langle\Sigma,C^C\rangle$. □

Proposition 5 Let $X_1,X_2,\ldots,X_k$ be the connected components of the graph $\langle\Sigma,C^C\rangle$. Then there always exists a path on the transition diagram of automaton $\langle M,q_0,\delta,F\rangle$ that joins the following states :

$\Sigma, \Sigma-X_1, \Sigma-X_1-X_2, \ldots, X_K, \emptyset$.

Proof. In order to build the path, we start from state Σ and repeatedly apply the transition function to symbols σ' ($\in \Sigma'$) such that the corresponding elements σ ($\in \Sigma$) belong to X_1. Proposition 4 guarantees that after a certain number of such transitions the automaton will enter the state $\Sigma - X_1$. Then we go on applying δ to symbols $\sigma' \in \Sigma'$ such that $\sigma \in X_2$. Again, after some transitions the automaton will enter the state $\Sigma - X_1 - X_2$, and so forth up to the state $\emptyset$. □

Now we give the following definitions:

- a finite sequence $(S_1, S_2, \ldots, S_r)$, where $S_i \in 2^{\Sigma}$ $\forall$ i, is called a *route* and will be denoted by R if $S_1 = \Sigma$, $S_i \neq S_j$ $\forall$ $i \neq j$, and there exists a path $(x_1, x_2, \ldots, x_r)$ on the transition diagram of $\langle M, q_0, \delta, F\rangle$ such that $\forall$ $i = 1, r-1$ $\delta(S_i, x_i) = S_{i+1}$. Note that $\forall$ $i = 1, r-1$ $x_i \in \Sigma'$;
- a route $R = (S_1, \ldots, S_r)$ is called a *main route*, and will be denoted by R^*, if it includes a subsequence $(S_{i_1}, S_{i_2}, \ldots, S_{i_{k+1}})$ such that $S_{i_1} = \Sigma$, $S_{i_j} = \Sigma - X_1 - X_2 - \ldots - X_{j-1}$ $\forall$ j, where $X_1, X_2, \ldots, X_k$ are the connected components of $\langle \Sigma, C^c \rangle$;
- we shall say that a string $z = z_1 \ldots z_n \in (\Sigma \cup \Sigma')^*$ *lies* on a route $R = (S_1, \ldots, S_m)$, $m \leq n$, if $(S_1, \ldots, S_m)$ is the sequence of *distinct* states entered by the automaton $\langle M, q_0, \delta, F\rangle$ when, starting from state q_0, the function δ is applied consecutively to the symbols $z_1, z_2, \ldots z_n$;
- for every route R, let $f_R(n)$ denote the number of strings of length n that lie on R, while $f_R(x)$ will represent the corresponding generating function.

Proposition 6 For every main route R^*, $f_{R^*}(n) = \theta(n^k\,|\Sigma|^n)$.

Proof. Let $R^* = (S_1, S_2, \ldots, S_m)$ be a main route. Then it is easily shown that

$$f_{(S_1,S_2)}(n) = \omega(S_1, S_2) \sum_{i=0,n-1} (l(S_1))^i (l(S_2))^{n-i-1};$$

so the corresponding generating function is

$$f_{(S_1,S_2)}(x) = \omega(S_1, S_2)\, x\, \{(1 - l(S_1)x)\,(1 - l(S_2)x)\}^{-1}.$$

Reasoning by induction we can determine the generating function of $\{f_{R^*}(n)\}$:

$$f_{R^*}(x) = A\, x^{m-1} \{\prod_{i=1,m} (1 - l(S_i)x)\}^{-1},$$

where A is constant. Since R^* is a main route, by propositions 4 and 5, the last expression can be written in the form

$$A \cdot x^{m-1} \cdot \{ (1-|\Sigma| x)^{k+1} \prod_{i=1,m-k-1} (1-a_i x) \}^{-1} ,$$

where $a_i < |\Sigma| \ \forall i$. So, applying lemma 1 to the last expression, we obtain the result.□

Proposition 7 Let R be a route and suppose it is not main; then, for every main route R^*, $f_R(n)=o(f_{R^*}(n))$.

Proof. Applying the same reasoning as in the previous proof, we can determine the generating function of $f_R(n)$. Then, applying lemma 1, we obtain : $f_R(n) = \theta(n^{T-1} |\Sigma|^n)$, where $1 \le T \le k$ and being T the number of states belonging to R which are represented by union of connected components in $\langle \Sigma, C^c \rangle$. Therefore the thesis follows from proposition 6.□

Now we are able to prove the main result of this section.

Theorem 2 $E(n) = \theta(n^k)$.

Proof. For every main route R^* we know that, by proposition 2,

$$f_{R^*}(n) \le f(n) \le \Sigma_R f_R(n) ;$$

applying the results of propositions 6 and 7, and recalling that the number of ruotes is constant, we can state

$$f(n) = \theta(n^k |\Sigma|^n) .$$

Since $E(n) = f(n)/|\Sigma|^n$, the thesis follows.□

5. Average analysis B

In this section we consider the following probability distribution on Σ^n: each string of length n has probability $p(x) = \{(\#[x]_C) \cdot t_n\}^{-1}$, where t_n is the number of traces of length n. This assumption is equivalent to consider each trace of length n as equiprobable. So the problem is reduced to the evaluation of $E_t(n) = \mathfrak{I}_n / t_n$, where $\mathfrak{I}_n = \Sigma_{|t|=n} \mathfrak{I}(t)$.

In order to evaluate the asymptotic expressions of $\mathfrak{I}_n$ and t_n, we recall some concepts of formal series theory.[8]

In the following, we only consider monoids with finite decompositions [8]. The set S(M) of formal series on the monoid M is defined as $\{\Sigma_{t \in M} g(t) \cdot t \ / \ g : M \rightarrow \mathbb{C}\}$, where $\mathbb{C}$ is the complex field. Representing the sum of series by + and the Cauchy product by ●, we know that $\langle S(M), +, \bullet \rangle$ is an integral domain. We shall abbreviate the expression of the generic series in the form $\Sigma_t g(t)\cdot t$.

Let $\mathcal{L}$ denote the set of cliques of $\langle\Sigma,C\rangle$. We consider the formal series $\xi = \Sigma_t t$, and the Moebius function $\mu = \Sigma_{B\in\mathcal{L}}(-)^{|B|} B$.

It is known from formal series theory that :

A) $\xi \bullet \mu = \varepsilon$ (where ε represents the unity in $S(F(\Sigma,C))$).

Let $\{x\}^*$ be the monoid of the strings composed only by one symbol. The function $\psi : F(\Sigma,C) \rightarrow \{x\}^*$ such that $\psi(t) = x^{|t|}$ is a surjective morphism between the two monoids; so the function $\psi^\circ : S(F(\Sigma,C)) \rightarrow S(\{x\}^*)$, where $\psi^\circ(\Sigma_t g(t)\cdot t) = \Sigma_n \{\Sigma_{|t|=n} g(t)\} x^n$, is a morphism between $S(F(\Sigma,C))$ and $S(\{x\}^*)$. Therefore we can write :

$\psi^\circ(\xi) = \Sigma_n t_n x^n$, $\psi^\circ(\mu) = \Sigma_{j=0,\alpha}(-)^j \Gamma_j x^j$, where α is the size of the maximum clique in $\langle\Sigma,C\rangle$ and Γ_j is the number of j-cliques. Moreover, since it is easily seen that $\xi^2 = \Sigma_t \mathfrak{I}(t)\cdot t$, we can state that

$$\psi^\circ(\xi^2) = \Sigma_n(\Sigma_{|t|=n} \mathfrak{I}(t))\, x^n = \Sigma_n \mathfrak{I}_n x^n .$$

Recalling that ψ° is a morphism, from relation A) it follows :

A') $\Sigma_n t_n x^n = \{\Sigma_{j=0,\alpha}(-)^j \Gamma_j x^j\}^{-1}$;

B') $\Sigma_n \mathfrak{I}_n x^n = \{\Sigma_{j=0,\alpha}(-)^j \Gamma_j x^j\}^{-2}$.

Also in this case, we have to evaluate two sequences the generating functions of which are known.

Theorem 3 Let Γ_j be the number of j-cliques in $\langle\Sigma,C\rangle$; assume that the polynomial $G(x) = \Sigma_{j=0,\alpha}(-)^j \Gamma_j x^j$ has only one root b of least modulus. Then $E_t(n) = \theta(n^h)$, where h is the multiplicity of b in G(x) .

Proof. By lemma 1 we can deduce from A') and B') that

$t_n = \theta(n^{h-1}(1/b)^n)$, and

$\mathfrak{I}_n = \theta(n^{2h-1}(1/b)^n)$.

We conclude that $E_t(n) = \mathfrak{I}_n / t_n = \theta(n^h)$.□

6. Examples and open problems

According to the previous results, the following table shows the time complexity (worst case, average case) of the procedure scheme on some concurrent alphabets.

$\langle\Sigma,C\rangle$	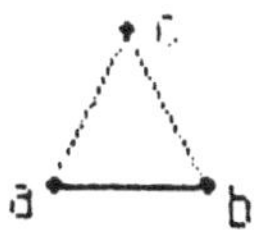		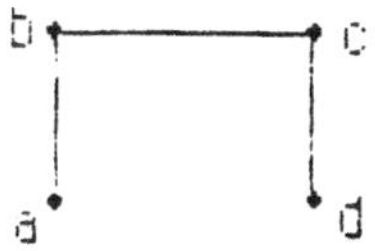	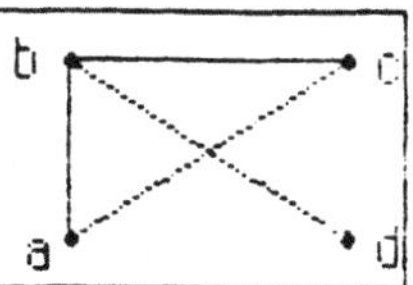
$G(x)$	$(1-x)^3$	$1-3x+2x^2$	$1-4x+3x^2$	$1-4x+4x^2-x^3$
$W(n)$	$\theta(n^3)$	$\theta(n^2)$	$\theta(n^2)$	$\theta(n^3)$
$E(n)$	$\theta(n^3)$	$\theta(n^2)$	$\theta(n)$	$\theta(n^2)$
$E_t(n)$	$\theta(n^3)$	$\theta(n)$	$\theta(n)$	$\theta(n)$

It is worth noting that for some concurrent alphabets $W(n)=\theta(E_t(n))=\theta(E(n))$, while, for other concurrent alphabets, $E_t(n)=o(E(n))$ and $E(n)=o(W(n))$. While trivially $E(n)=O(W(n))$ and $E_t(n)=O(W(n))$, we also conjecture that $E_t(n)=O(E(n))$.

Another interesting example is obtained when the concurrent alphabet is a random graph. Let p be a constant such that $0<p<1$; let $G(N,p)$ denote a random graph with N nodes such that each edge occurs with probability p independently of all other edges. Let $\alpha(N,p)$ denote the size of the maximum clique in $G(N,p)$. The evolution of a random graph, when N tends to ∞, has been studied in the literature. In particular it is known that the following events occur with probability 1 :

- $\alpha(N,p)$ is asymptotic to $2\log_{1/p}N$, for N tending to ∞ ([6]);
- there exists N_0 such that $\forall N > N_0$ $G(N,p)$ is connected ([5]).

In other words we are almost sure that, for enough large N, $G(N,p)$ is connected and $\alpha(N,p)$ grows as $2\log_{1/p}N$.

Let us suppose that the concurrent alphabet $\langle\Sigma,C\rangle$ is a random graph $G(|\Sigma|,p)$. So, when $|\Sigma|$ is large, we can state that, with probability close to1, the time required by the scheme in the worst case is $\theta(n^{2\log(|\Sigma|)+\eta})$, where $\lim_{|\Sigma|\to\infty}\eta/\log(|\Sigma|) = 0$, and the average time in the case of equidistributed strings is $\theta(n)$. This fact tells us that, for large "random" alphabet, the average behaviour is remarkably better than the worst case behaviour.

It would be interesting to know what the average time complexity is in the case of equidistributed traces when the concurrent alphabet is supposed a random graph.

References

[1] Bertoni A., Brambilla M.,Mauri G., Sabadini N., "An application of the theory of free partially commutative monoids: asymptotic densities of trace languages", Lect. Not. Comp. Sci. 118, Springer Berlin, 1981,pp. 205-215.

[2] Bertoni A., Mauri G., Sabadini N., "Equivalence and membership problems for regular trace languages", Proc. 9th ICALP, Lect. Not. Comp. Sci. 140 (Nielsen-Schmidt ed's), pp. 61-71, 1982.

[3] Bertoni A., Mauri G., Sabadini N., "Representation of prefixes of a trace and membership problem for context free trace languages", Int. Rep. Ist. Cib. Univ. Milano, 1985.

[4] Cartier P., Foata D.,"Problemes combinatoires de commutation et rearrangements", Lect. Not. in Math. 85, Springer , Berlin, 1969.

[5] Erdos P., Renyi A., "On random graphs", Publ. Math. Debrecen 6 (1959) pp. 290-297.

[6] Grimmet G.R., McDiarmid C.J.H., "On colouring random graph", Math. Proc. Camb. Philos. Soc. 77 (1975), pp.313-324.

[7] Janicki R., "Synthesis of concurrent schemes", Lect. Not. in Comp. Sci. 64, Springer Berlin (1978), pp. 298-307.

[8] Lallement G., "Semigroups and combinatorial applications", J. Wiley and sons, 1979.

[9] Mazurkiewicz A., "Concurrent program schemes and their interpretations", DAIMI, PB 78, Aarhus University, 1977.

[10] Rozenberg G., Verraedt R., "Subset languages of Petri nets", Proc. 3rd European workshop on applications and theory of Petri nets, Varenna Italy, 1982.

[11] Szijàrtò M., "Trace languages and closure operations", Automata Theoretic Letters, 1979/2, Dep. of Num. and Comp. Math., Eotvos Univ.,Budapest 1979.

AUTHORS INDEX

Vol. 270: E. Börger (Ed.), Computation Theory and Logic. IX, 442 pages. 1987.

Vol. 271: D. Snyers, A. Thayse, From Logic Design to Logic Programming. IV, 125 pages. 1987.

Vol. 272: P. Treleaven, M. Vanneschi (Eds.), Future Parallel Computers. Proceedings, 1986. V, 492 pages. 1987.

Vol. 273: J.S. Royer, A Connotational Theory of Program Structure. V, 186 pages. 1987.

Vol. 274: G. Kahn (Ed.), Functional Programming Languages and Computer Architecture. Proceedings. VI, 470 pages. 1987.

Vol. 275: A.N. Habermann, U. Montanari (Eds.), System Development and Ada. Proceedings, 1986. V, 305 pages. 1987.

Vol. 276: J. Bézivin, J.-M. Hullot, P. Cointe, H. Lieberman (Eds.), ECOOP '87. European Conference on Object-Oriented Programming. Proceedings. VI, 273 pages. 1987.

Vol. 277: B. Benninghofen, S. Kemmerich, M.M. Richter, Systems of Reductions. X, 265 pages. 1987.

Vol. 278: L. Budach, R.G. Bukharajev, O.B. Lupanov (Eds.), Fundamentals of Computation Theory. Proceedings, 1987. XIV, 505 pages. 1987.

Vol. 279: J.H. Fasel, R.M. Keller (Eds.), Graph Reduction. Proceedings, 1986. XVI, 450 pages. 1987.

Vol. 280: M. Venturini Zilli (Ed.), Mathematical Models for the Semantics of Parallelism. Proceedings, 1986. V, 231 pages. 1987.

Vol. 281: A. Kelemenová, J. Kelemen (Eds.), Trends, Techniques, and Problems in Theoretical Computer Science. Proceedings, 1986. VI, 213 pages. 1987.

Vol. 282: P. Gorny, M.J. Tauber (Eds.), Visualization in Programming. Proceedings, 1986. VII, 210 pages. 1987.

Vol. 283: D.H. Pitt, A. Poigné, D.E. Rydeheard (Eds.), Category Theory and Computer Science. Proceedings, 1987. V, 300 pages. 1987.

Vol. 284: A. Kündig, R.E. Bührer, J. Dähler (Eds.), Embedded Systems. Proceedings, 1986. V, 207 pages. 1987.

Vol. 285: C. Delgado Kloos, Semantics of Digital Circuits. IX, 124 pages. 1987.

Vol. 286: B. Bouchon, R.R. Yager (Eds.), Uncertainty in Knowledge-Based Systems. Proceedings, 1986. VII, 405 pages. 1987.

Vol. 287: K.V. Nori (Ed.), Foundations of Software Technology and Theoretical Computer Science. Proceedings, 1987. IX, 540 pages. 1987.

Vol. 288: A. Blikle, MetaSoft Primer. XIII, 140 pages. 1987.

Vol. 289: H.K. Nichols, D. Simpson (Eds.), ESEC '87. 1st European Software Engineering Conference. Proceedings, 1987. XII, 404 pages. 1987.

Vol. 290: T.X. Bui, Co-oP A Group Decision Support System for Cooperative Multiple Criteria Group Decision Making. XIII, 250 pages. 1987.

Vol. 291: H. Ehrig, M. Nagl, G. Rozenberg, A. Rosenfeld (Eds.), Graph-Grammars and Their Application to Computer Science. VIII, 609 pages. 1987.

Vol. 292: The Munich Project CIP. Volume II: The Program Transformation System CIP-S. By the CIP System Group. VIII, 522 pages. 1987.

Vol. 293: C. Pomerance (Ed.), Advances in Cryptology – CRYPTO '87. Proceedings. X, 463 pages. 1988.

Vol. 294: R. Cori, M. Wirsing (Eds.), STACS 88. Proceedings, 1988. IX, 404 pages. 1988.

Vol. 295: R. Dierstein, D. Müller-Wichards, H.-M. Wacker (Eds.), Parallel Computing in Science and Engineering. Proceedings, 1987. V, 185 pages. 1988.

Vol. 296: R. Janßen (Ed.), Trends in Computer Algebra. Proceedings, 1987. V, 197 pages. 1988.

Vol. 297: E.N. Houstis, T.S. Papatheodorou, C.D. Polychronopoulos (Eds.), Supercomputing. Proceedings, 1987. X, 1093 pages. 1988.

Vol. 298: M. Main, A. Melton, M. Mislove, D. Schmidt (Eds.), Mathematical Foundations of Programming Language Semantics. Proceedings, 1987. VIII, 637 pages. 1988.

Vol. 299: M. Dauchet, M. Nivat (Eds.), CAAP '88. Proceedings, 1988. VI, 304 pages. 1988.

Vol. 300: H. Ganzinger (Ed.), ESOP '88. Proceedings, 1988. VI, 381 pages. 1988.

Vol. 301: J. Kittler (Ed.), Pattern Recognition. Proceedings, 1988. VII, 668 pages. 1988.

Vol. 302: D.M. Yellin, Attribute Grammar Inversion and Source-to-source Translation. VIII, 176 pages. 1988.

Vol. 303: J.W. Schmidt, S. Ceri, M. Missikoff (Eds.), Advances in Database Technology – EDBT '88. X, 620 pages. 1988.

Vol. 304: W.L. Price, D. Chaum (Eds.), Advances in Cryptology – EUROCRYPT '87. Proceedings, 1987. VII, 314 pages. 1988.

Vol. 305: J. Biskup, J. Demetrovics, J. Paredaens, B. Thalheim (Eds.), MFDBS 87. Proceedings, 1987. V, 249 pages. 1988.

Vol. 306: M. Boscarol, L. Carlucci Aiello, G. Levi (Eds.), Foundations of Logic and Functional Programming. Proceedings, 1986. V, 218 pages. 1988.

Vol. 307: Th. Beth, M. Clausen (Eds.), Applicable Algebra, Error-Correcting Codes, Combinatorics and Computer Algebra. Proceedings, 1986. VI, 215 pages. 1988.